# Multi-scale Simulation of Metallic Materials (2nd Edition)

# Multi-scale Simulation of Metallic Materials (2nd Edition)

Guest Editor

**Changming Fang**

MDPI

Basel • Beijing • Wuhan • Barcelona • Belgrade • Novi Sad • Cluj • Manchester

*Guest Editor*
Changming Fang
Brunel Centre for Advanced
Solidification Technology
(BCAST)
Brunel University London
Uxbridge
UK

*Editorial Office*
MDPI AG
Grosspeteranlage 5
4052 Basel, Switzerland

This is a reprint of the Special Issue, published open access by the journal *Metals* (ISSN 2075-4701), freely accessible at: https://www.mdpi.com/journal/metals/special_issues/T580121GW3.

For citation purposes, cite each article independently as indicated on the article page online and as indicated below:

| Lastname, A.A.; Lastname, B.B. Article Title. *Journal Name* **Year**, *Volume Number*, Page Range. |
|---|

**ISBN 978-3-7258-7368-5 (Hbk)**
**ISBN 978-3-7258-7369-2 (PDF)**
**https://doi.org/10.3390/books978-3-7258-7369-2**

# Contents

# About the Editor

**Changming Fang**

Changming Fang obtained the doctoral degree of philosophy with a specialization in Materials Science, in the Solid State Chemistry Lab, National University of Groningen (RijksUniversiteit Groningen), the Netherlands (1996). He has worked in world-wide institutes including Ångstrom Lab., Uppsala University, Sweden; Computational Materials Physics, University of Vienna, Austria; Materials Innovation Institute (M2I), Delft University of Technology and Debye Institute for Nanomaterials Science, Utrecht University; and Chemical Engineering and Chemistry Lab., Eindhoven University of Technology, Electronic Structure Lab., Radboud University Nijmegen, the Netherlands; University of Rennes 1, France; and Arkansas University, USA. In the last ten years (2016–2025), he has worked as a senior researcher in the Brunel Centre for Advanced Solidification Technology (BCAST), Brunel University London, the United Kingdom. Dr. Fang has journeyed in various fields and topics (see the attached list of publications) including high-strength and ductile steels (TRIP, dual-phase steels, etc.); light metals (Mg, Al) alloys and intermetallic compounds and precipitates; liquid–metal/solid interfaces which relate to (pre)nucleation and solidification processes of metals/alloys; heterojunctions for solar cells, photon detection devices, etc.; the new silicon oxynitride based luminescence materials; low-dimensional materials, including transition metal dichalcogenides, the intercalates and misfit layered compounds, and novel nanosheets and nanotubes, etc.; the bulk properties and surface structure, energetics of various oxides and nitrides, and (oxy)nitrides using ab initio tools combined with the experiments; novel functional materials (e.g., half-metallic materials), etc. Moreover, Changming Fang had about ten years of experimental experience. He has published over 160 referenced papers (Changming Fang—Google Scholar).

metals MDPI

Editorial

# Multi-Scale Simulation of Metallic Materials (2nd Edition)

**Changming Fang**

Brunel Centre for Advanced Solidification Technology (BCAST), Brunel University London,
Uxbridge UB8 3PH, UK; changming.fang@brunel.ac.uk

## 1. Introduction

Metallic materials are some of the most important engineering materials. Current developments in our society are leading to an increasing demand for diverse novel metallic materials with desirable properties. Furthermore, there is an urgent need for these metallic materials to be manufactured in an environmentally friendly way with low energy costs, in addition to a demand for the sustainable recycling of metallic scrap/waste which contains a rich variety of impurities [1]. To reach such goals, the use of both experimental and theoretical tools to obtain comprehensive knowledge regarding metallic materials on scales ranging from the atomic-, micro-, meso- to macroscopic-level is essential.

Great efforts have been made to design novel ferrous metals and light metals including Al and Mg from, e.g., metallic scraps [2]. Moreover, improving the mechanical performance of metallic materials is a big challenge for their application. Among the major alloy-strengthening mechanisms, the uniform dispersion of nanoscale secondary phase precipitates is particularly effective. Ceramic materials have been added into metal matrices to produce nanocomposites that exhibit excellent mechanical performance [3,4]. In addition to this, new types of metallic materials, such as multicomponent high-entropy alloys [5], have been intensively explored.

Recent progress in computational capability and modeling techniques has promoted the application of multiscale simulation techniques in metallic materials [6]. Multiscale simulations which combine existing and emerging methods are currently being employed to incorporate a wide range of time and space scales that are inherent to various disciplines. Moreover, combining experimental observations and multiscale modeling provides us with a comprehensive understanding of the engineering processes and physical properties of metallic products.

Considering this, this Special Issue aims to improve our understanding of the structural, microstructural, and physical properties of complex metallic materials via multiscale approaches, including thermodynamics, finite element methods, and ab initio molecular dynamics simulations.

## 2. Overview of the Contributions

Zabojszcza and colleagues contributed a paper entitled 'Verification of Numeric Models of Steel Bar Coverings Using Experimental Tests—Preliminary Study' (Contribution 1). Construction failures causing damage in load-bearing structures made of steel are considered some of the most severe structural failures, not only in terms of material loss but also in terms of potential human casualties. The mitigation of these occurrences is decisive in the design of steel roofs. This work aimed to verify and assess the accuracy of a macroscopic numerical model of a metal bar roof by conducting experimental studies. The structure covered by the metal bar roof was characterized by a strongly nonlinear static response;

ignoring these nonlinear effects leads to large computational errors. The authors performed a series of repeatable experimental tests on the structure model to determine the path of static equilibrium and the form of stability loss of the steel covering using the relation between the load and the displacements of nodes. The displacements of the nodes were verified using precise triangulation laser sensors and electronic sensors. For a low-rise bar covering, the inclusion of imperfections is particularly important, which significantly affects the local and overall stability of the structure. The obtained information helps us to understand the accuracy and reliability of numerical models. In turn, it contributes to the development of improved methods for the analysis of metal bar roof structures in future works.

Quan et al. presented their work 'A Characterization of the Powder Yield Behaviors During a Hot Isostatic Pressing Process' (Contribution 2). It has been well-established that temperature plays a crucial role in determining material properties during the hot isostatic pressing of powders. It is therefore essential to account for its effect on the yield model parameters in order to accurately and reliably describe the densification evolution of powders. The authors developed a mesoscopic, particle-scale, high-temperature uniaxial compression model based on the discrete element method. The predicted relations between strain and stress and corresponding densities in the high-temperature uniaxial compression simulations were verified via experiments. The authors analyzed the strain evolution during the uniaxial compression process and discovered the relationship between the parameters of the Shima–Oyane model and temperature. Based on the obtained parameters, they created a yield stress map for the nickel-based alloy to act as an example. The accuracy/reliability of the model was verified by comparing the experimental results with the finite element method (FEM) simulations. This study contributes to a more reliable prediction of densification behavior in the thermally driven isostatic pressing of metallic materials.

Jacobson and colleagues provided their recent work entitled 'Extending Density Phase-Field Simulations to Dynamic Regimes' (Contribution 3). Grain boundaries play a crucial role in determining the physical properties of metallic materials. Density-based phase-field (DPF) methods have emerged as a technique for simulating grain boundary thermodynamics and kinetics. Compared to the classical phase-field, DPF gives a more physical description of the grain boundary structure and chemistry. It bridges the gap between CALPHAD databases and atomistic simulations and it thus has potential applications in grain boundary and segregation engineering. The authors generated a generic DPF free energy functional which was used to carry out a series of equilibrium and dynamic simulations of grain boundaries, producing relations such as grain boundary width vs. gradient energy coefficient, grain boundary velocity vs. applied driving force, and spherical grain radius vs. time. These relations are compared with analytical solutions and the behavior of physical grain boundaries and are used to assess the validity of the coupled DPF model. A good agreement was reached between the tested quantities and established theories of grain boundary behavior, which indicated that coupling density fields with traditional order parameters in DPF simulations is an excellent approach for making DPF simulations dynamic and for improving the simulation's performance and accuracy.

Andersson and Lundbäck provided their work 'Modeling the Evolution of Grain Texture during Solidification of Laser-Based Powder Bed Fusion Manufactured Alloy 625 Using a Cellular Automata Finite Element Model' (Contribution 4). In additive manufacturing, the materials are added through a layer-by-layer method. Then, by subsequently melting the added material with a heat source, it is fused into a solid structure. The grain texture of the as-printed material evolves during the laser-based powder bed fusion (PBF-LB) process. The resulting mechanical properties depend on the obtained grain texture and the chosen


https://doi.org/10.3390/met16030276

process parameters, including scan velocity and laser power. The researchers developed a coupled 2D Cellular Automata and Finite Element model (2D CA-FE) to predict the evolution of the grain texture during solidification of the nickel-based superalloy 625 produced by PBF-LB. The finite element (FE) model predicted the temperature history of the building and the CA model made predictions of nucleation and grain growth based on the temperature history. The 2D CA-FE model captured the solidification behavior observed in PBF-LB, such as competitive grain growth plus equiaxed and columnar grain growth. Three different nucleation densities for heterogeneous nucleation were investigated. The optimized nucleation density was found to give the best result as compared to the existing EBSD data in the literature. With the selected nucleation density, the predicted aspect ratio and grain size distribution of the simulated grain texture agreed well with the observed textures from EBSD in the literature.

Bin Abd Rahim et al. published the paper 'Assessing Fatigue Life Cycles of Material X10CrMoVNb9-1 through a Combination of Experimental and Finite Element Analysis' (Contribution 5). The authors combined experiments and theoretical modeling for investigating fatigue crack initiation and propagation of the material X10CrMoVNb9-1 (P91) steel under cyclic loading at room temperature. The Voronoi tessellation method was implemented to generate an artificial microstructure model at the microstructure level; following this, the finite element (FE) method was applied to identify different stress distributions. The stress distributions for multiple artificial microstructures were analyzed by using the physically based Tanaka–Mura model to estimate the number of cycles for crack initiation. For the prediction of macro-scale and long-term crack formation, the Paris law was utilized in this research. Experimental work on fatigue life with this material was performed, and good agreement was found with the results obtained in FE modeling. The number of cycles for fatigue crack propagation attains up to a maximum of 40% of the final fatigue lifetime with a typical value of 15% in many cases. This physically based two-scale technique significantly advances fatigue research, particularly in power plants, and paves the way for rapid and low-cost virtual material analysis and fatigue resistance analysis in the context of environmental fatigue applications.

Zeng and colleagues presented their results 'Crystal Plasticity Finite Element Simulation of Tensile Fracture of 316L Stainless Steel Produced by Selective Laser Melting' (Contribution 6). In the current work, the authors employed a combination of experimental observations and numeric modeling approaches to investigate the elective laser melting (SLM) of 316L stainless steel. The influence of building direction and grain boundary strength on the fracture parameters of SLM 316L stainless steel was studied using electron backscatter diffraction (EBSD) experiments which characterized the microstructure of the specimens. A representative volume element (RVE) model reflecting the microstructure of SLM 316L stainless steel was established based on a combination of the crystal plastic finite element method (CPFEM) and UMAT subroutine technology. The crystal plasticity parameters were obtained by means of analyzing the results of tensile tests. Cohesive elements were employed and inserted at the grain boundaries of the polycrystalline RVE to simulate the intergranular fracture behavior of SLM 316L stainless steel under uniaxial tensile loading. The damage and fracture mechanisms of the material at the microscale were analyzed. The simulated tensile stress–strain curves were in good agreement with the experimental results; hence, the combined CPFEM model is suitable for characterizing the mechanical response and fracture behavior of the steel. The results revealed that cracks initiate at stress concentration sites and propagate along grain boundaries with increasing external load, ultimately leading to rupture. Moreover, the building direction influences the location of microcracks and their propagation significantly.


https://doi.org/10.3390/met16030276

Fang and Fan present their work 'Crystal Chemistry at Interfaces Between Liquid Al and Polar SiC{0001} Substrates' (Contribution 7) in the Special Issue. Silicon carbide (SiC) has been widely added into light metals, e.g., Al and Mg, to enhance their mechanical performance and corrosion resistance. SiC particle-reinforced metal matrix composites (SiC-MMCs) exhibit low weight/volume ratios, high strength/hardness, high corrosion resistance, and thermal stability. They have potential applications in aerospace, automobiles, and other specialized equipment. The macro-mechanical properties of Al/SiC composites depend on the local structures and chemical interactions at the Al/SiC interfaces at the atomic level. Moreover, the added SiC particles may act as potential nucleation sites during solidification. The present authors investigated local atomic ordering and chemical interactions at the interfaces between liquid Al (Al(l) in short) and the polar SiC substrates using ab initio molecular dynamics (AIMD) methods. The simulations reveal a rich variety of interfacial interactions. Charge transfer occurs from Al(l) to C-terminating atoms ($\Delta q = 0.3$ e/Alon average), while chemical bonding between interfacial Si and Al(l) atoms is more covalent with a minor charge transfer of $\Delta q = 0.04$ e/Al. The prenucleation at both interfaces is moderate with three to four recognizable layers. The information obtained here helps increase understanding of the interfacial interactions at Al/SiC at the atomic level and the related macro-mechanical properties, which is helpful in designing novel SiC-MMC materials with desirable properties and optimizing related manufacturing and machining processes.

Führer et al. published the paper 'Experimental Investigation of the Interplay Between Al-, B-, and Ti-Nitrides in Microalloyed Steel and Thermodynamic Analysis' (Contribution 8). The primary objective of this study was to accurately describe the microstructural evolution in the Fe-Al-B-Ti-C-N system using the Calphad approach, with special emphasis on correctly predicting the dissolution temperatures of nitrides. The authors built a multicomponent database through the incorporation of available binary and ternary descriptions, employing the Calphad method. The experimental findings regarding the solvus temperature of the involved nitrides were employed to validate the accuracy of the thermodynamic database. The findings offered a comprehensive understanding of the relative phase stabilities and the associated interplay among the involved elements Al, B, and Ti in the Fe-rich corner of the system. The type and size distributions of the stable nitrides in micro-alloyed steel have been demonstrated to exert a substantial influence on the properties of the material, thereby rendering accurate predictions of phase stabilities of considerable relevance.

## 3. Summary and Outlook

The second edition of this Special Issue of *Metals* was supported by various research groups and a final book of eight high-quality peer-reviewed articles (Multi-scale Simulation of Metallic Materials (2nd Edition) | Metals | MDPI). The success of this special issue will be followed by an. other special issue (Advances in the Study of Metal Crystals) to accept global contributions from the metal community.

**Funding:** This research received no external funding.

**Conflicts of Interest:** The author declares no conflicts of interest.



https://doi.org/10.3390/met16030276

**List of Contributions:**

1. Zabojszcza, P.; Radoń-Kobus, K.; Kossakowski, P.G. Verification of Numerical Models of Steel Bar Coverings Using Experimental Tests—Preliminary Study. *Metals* **2024**, *14*, 1319. https://doi.org/10.3390/met14121319.
2. Quan, G.Z.; Ran, W.J.; Dai, W.W.; Jiang, Q.; Yu, Y.Z.; Zhang, Y. A Characterization of the Powder Yield Behaviors During a Hot Isostatic Pressing Process. *Metals* **2025**, *15*, 752. https://doi.org/10.3390/met15070752.
3. Jackson, D.; Kamachali, R.D.; Thompson, G.B. Extending density phase-field simulations to dynamic regimes. *Metals* **2023**, *13*, 1497. https://doi.org/10.3390/met13081497.
4. Andersson, C.; Lundbäck, A. Modeling the Evolution of Grain Texture during Solidification of Laser-Based Powder Bed Fusion Manufactured Alloy 625 Using a Cellular Automata Finite Element Model. *Metals* **2023**, *13*, 1846. https://doi.org/10.3390/met13111846.
5. Bin Abd Rahim, M.R.; Schmauder, S.; Manurung, Y.H.P.; Binkele, P.; Dusza, J.; Csanádi, T.; Ahmad, M.I.M.; Muhd Faiz Mat, M.F.; Dogahe, K.J. Assessing Fatigue Life Cycles of Material X10CrMoVNb9-1 through a Combination of Experimental and Finite Element Analysis. *Metals* **2023**, *13*, 1947. https://doi.org/10.3390/met13121947.
6. Zeng, G.W.; Huang, Z.Y.; Deng, B.; Ge, R. Crystal Plasticity Finite Element Simulation of Tensile Fracture of 316L Stainless Steel Produced by Selective Laser Melting. *Metals* **2025**, *15*, 567. https://doi.org/10.3390/met15050567.
7. Fang, C.M. Fan, Z. Crystal Chemistry at Interfaces Between Liquid Al and Polar SiC{0001} Substrates. *Metals* **2024**, *14*, 1258. https://doi.org/10.3390/met14111258.
8. Führer, M.; Zamberger, S.; Seubert, C.; Povoden-Karadeniz, E. Experimental Investigation of the Interplay Between Al-, B-, and Ti-Nitrides in Microalloyed Steel and Thermodynamic Analysis. *Metals* **2025**, *15*, 705. https://doi.org/10.3390/met15070705.

## References


1. Reuter, M.A.; van Schaik, A.; Gutzmer, J.; Bartie, N.; Abadías-Liamas, A. Challenges of the circular economy: A material, Metallurgical, and Product design perspective. *Ann. Rev. Mater. Res.* **2019**, *49*, 253–274. [CrossRef]
2. Rane, K.; Nayak, K.; Date, P.; Tirumalai, S. Solid state recycling of metal scrap from manufacturing net-shaped parts. *Int. J. Met. Mater.* **2025**, *1*, 21–38.
3. Singh, A.; Singh, J.; Sinha, M.K. Ferrous-metal matrix composites: A review on status, scope and challenges. *Int. J. Interact. Des. Manuf.* **2023**, *17*, 2807–2829. [CrossRef]
4. Liu, Y.Y.; Liu, Z.Q.; Liu, Z.Y.; Zhou, W.H.; Yu, S.; Xiao, B.L.; Ma, Z.Y.; Zhang, Z.F.; Ritchie, R.O. Nature inspires new high-performance metal composites. *Interdis. Mater.* **2025**, *4*, 502–507. [CrossRef]
5. Cantor, B. Multicomponent high-entropy Cantor alloys. *Porg. Mater. Sci.* **2021**, *120*, 100754. [CrossRef]
6. Fish, J.; Wagner, G.J.; Keten, S. Mesoscopic and multiple modelling in materials. *Nat. Mater.* **2021**, *20*, 774–786. [CrossRef] [PubMed]






https://doi.org/10.3390/met16030276

*metals*

*Article*

# A Characterization of the Powder Yield Behaviors During a Hot Isostatic Pressing Process

**Guozheng Quan [1,2,*], Wenjing Ran [1], Weiwei Dai [2], Qian Jiang [2], Yanze Yu [1] and Yu Zhang [1]**

[1] Chongqing Key Laboratory of Advanced Mold Intelligent Manufacturing, School of Material Science and Engineering, Chongqing University, Chongqing 400044, China

[2] Jiangsu Advanced Manufacturing Engineering Technology Research Center for Marine Power System Parts, COSCO Shipping Marine Equipment & Spares (Nanjing) Co., Ltd., Nanjing 211121, China; dai.weiwei@coscoshipping.com (W.D.); jiang.qian1@coscoshipping.com (Q.J.)

* Correspondence: quanguozheng@cqu.edu.cn; Tel.: +86-15922900904

**Abstract:** The constitutive model significantly influences the accuracy of predicting the complex rheological behavior of hot isostatically pressed powders. The temperature plays a crucial role in determining material properties during hot isostatic pressing (HIP), making it essential to account for its effect on the yield model parameters to more accurately describe the densification evolution of powders. In this study, HIP experiments were conducted using two different process schemes, and the shrinkage deformation of the envelope under each scheme was analyzed. High-temperature uniaxial compression experiments were performed on HIP samples with varying densities to analyze and characterize the stress–strain response of the powder during HIP. A mesoscopic particle-scale high-temperature uniaxial compression model was developed based on the discrete element method (DEM), and the strain and stress values corresponding to different densities in the high-temperature uniaxial compression simulations were validated through experimental comparison. The strain evolution during the uniaxial compression process was analyzed, and the relationship between the parameters of the Shima–Oyane model and the temperature was established, leading to the development of a temperature-compensated Shima–Oyane model. Based on the obtained parameters at various densities and temperatures, a yield stress map for the nickel-based alloy was constructed. The accuracy of this model was verified by comparing experimental results with finite element method (FEM) simulations. The findings of this study contribute to a more precise prediction of densification behavior in thermally driven isostatic pressing.

**Keywords:** hot isostatic pressing; nickel-based alloy; powder particle; densification

## 1. Introduction

Hot isostatic pressing (HIP) is an advanced materials-processing technology. It consolidates powders at high temperatures and pressures to produce fully dense, near-net-shape components. Due to its ability to manufacture isotropic components with excellent mechanical properties, HIP is widely utilized in critical aerospace, biomedical, and energy industries [1–3]. However, the component performance is governed by the complex interplay of thermos–mechanical phenomena, including time–temperature–pressure dependencies, viscoplasticity deformation, and diffusion-driven pore closure [4,5]. These multiscale, interdependent mechanisms occur nonlinearly across varying stages of densification, complicating predictive modeling and optimization. Therefore, different constitutive equations are used to simulate the flow behavior and densification mechanisms of metal powders during HIP process, such as the power-law creep equation [6,7], Drucker-Prager model [8], the Can-Clay model [8], the Perzyna model [9,10], and the Shima–Oyane model [11,12]. It is worth noting that, although the Shima–Oyane model does not account for key mechanisms in HIP such as diffusional creep, grain boundary diffusion, and pore surface tension effects, it has nonetheless become a widely adopted approach for modeling densification, owing to its broad applicability across a wide density range and its compatibility with various alloy systems. Wang et al. [12] established a HIP densification map using the Shima–Oyane model. Liu's group [13] utilized the Perzyna model to investigate titanium alloy compacts' high-temperature plasticity and creep coupling behavior. Although the Shima–Oyane model is applied to most materials and provides highly accurate predictions across a wide density range, the stresses in these materials are temperature-dependent. Based on the subsequent simulation data and fitting results, it can be seen that the parameter $f'$ in the Shima–Oyane model exhibits an extremely strong temperature dependence. To accurately characterize the density evolution behavior of powders during HIP, the effect of the temperature on the yielding model parameters must be considered.

Since the powder materials cannot be regarded as standard volumetrically compressible continua, significant strain, shrinkage, and deformation occur during the HIP process [14,15]. The continuum model is only suitable for describing the overall shrinkage and deformation of the powder during the HIP process. However, it has limitations in accurately capturing the individual powder particles' motion and deformation behavior. Coupling the finite element method (FEM) with the discrete element method (DEM) enables representation of the powder particle system as a solid continuum deformable body, thereby achieving an accurate description of the powder flow, deformation, and contact behavior. At present, this coupled approach has been successfully applied in the analysis of powder densification mechanisms [16–20], the construction of yield surfaces [21,22], the quantification of friction effects [23,24], and the simulation of compaction processes [25,26]. Elguezabal et al. [27] utilized the model to predict mechanical behavior at the mesoscopic particle scale. Li et al. [28] established various initial packing structures and investigated different densification mechanisms for three powders: single-particle powder, binary powder, and powder with a normal particle size distribution. Zou et al. [29] conducted a numerical reproduction of tungsten powder and examined the effects of the HIP process parameters on its compaction behavior. To this end, it is essential to establish a cross-scale correlation model linking macroscopic mechanical properties with microscopic particle dynamics to elucidate the densification mechanism of powders during HIP, thereby providing theoretical support for process optimization and improving product performance.

In this study, the high-temperature uniaxial compression behavior of powder particles was accurately simulated using a cross-scale coupled FEM-DEM approach, through which a temperature-compensated Shima–Oyane-based model was developed. Firstly, a random stacking structure of Inconel 625 powder was generated using the DEM. Subsequently, a finite element model was introduced for a numerical simulation and analysis. Two different HIP experimental schemes were conducted to evaluate the shrinkage behavior of HIP parts under varying process conditions. High-temperature uniaxial compression experiments and simulations were carried out on HIP parts with different initial densities to systematically investigate and characterize their stress–strain responses during hot extrusion. The simulation results were rigorously validated against experimental data, confirming the accuracy and reliability of the established high-temperature uniaxial compression model. A quantitative relationship between Shima–Oyane model parameters and temperature was identified through integrated simulation and experimental analyses, resulting in the formulation of a temperature-compensated Shima–Oyane constitutive model. This model was implemented into finite element simulations, and its predictive accuracy was validated through detailed comparisons with experimental observations. Introducing the temperature-compensated Shima–Oyane model significantly deepens the understanding of the powder deformation behavior under HIP conditions and offers robust theoretical insights and guidance for the multi-objective optimization of process parameters.

## 2. A Solution Approach to Powder Yield Model During HIP Process

### 2.1. Improved Shima–Oyane Model: Its Solution Approach

Plastic deformation and substantial volume shrinkage occur during the powder HIP process. Since the conventional von Mises yield criterion cannot predict the yield behavior under volume changes, it cannot meet the numerical simulation needs of powder HIP. Therefore, considering the volume change, flow stress, and hydrostatic pressure during powder deformation, Shima et al. [30,31] introduced the effect of hydrostatic pressure and developed an improved yield criterion, from which the effect of volume shrinkage can be predicted. The formula of the yield criterion is as follows [30,31]:

$$F = \sqrt{\frac{1}{2}\left[(\sigma_1-\sigma_2)^2+(\sigma_2-\sigma_3)^2+(\sigma_1-\sigma_3)^2\right]+\left(\frac{\sigma_m}{f}\right)^2} \tag{1}$$

where $F$ is a parameter related to the yield strength of material; $\sigma_1$, $\sigma_2$, and $\sigma_3$ are the first, second, and third principal stresses, respectively, in the von Mises yield criterion; $\sigma_m$ is the hydrostatic pressure; $f$ represents the influence of hydrostatic pressure on the yield strength of porous materials. Since $F$ is related to the yield strength $\overline{\sigma_0}$ of the matrix material of the porous material, Equation (1) can be derived in the following form:

$$f'\overline{\sigma}_0 = \sqrt{\frac{1}{2}\left[(\sigma_1-\sigma_2)^2+(\sigma_2-\sigma_3)^2+(\sigma_1-\sigma_3)^2\right]+\left(\frac{\sigma_m}{f}\right)^2} \tag{2}$$

$$f'\overline{\sigma}_0 = \sqrt{\frac{1}{2}\left[(\sigma_1-\sigma_2)^2+(\sigma_2-\sigma_3)^2+(\sigma_1-\sigma_3)^2\right]+\left(\frac{\sigma_m}{f}\right)^2}$$

where $f'$ denotes the ratio of the applied stress on the porous material to the equivalent stress sustained by the fully dense matrix. It is also expressed as a function of the material's relative density. The revised form of the Shima–Oyane model is as follows:

$$F = \frac{1}{f'}\left(\frac{3}{2}\sigma_y\sigma_d+\frac{\sigma_m^2}{f^2}\right)^{0.5}-\sigma_y \tag{3}$$

where $F$ represents the external force; $\sigma_d$ represents the deviatoric stress tensor; $\sigma_m$ indicates the hydrostatic pressure; $\sigma_y$ represents the uniaxial yield stress. In the Shima model, the parameters $f$ and $f'$ directly govern the plastic deformation behavior during the densification of alloy powders. Under uniaxial pressure conditions, the principal stresses satisfy $\sigma_2 = \sigma_3$, $\sigma_m = \sigma_1/3$, from which the corresponding yield condition can be derived:

$$f = \frac{\sqrt{2}}{3}\left(\frac{d\varepsilon_1-d\varepsilon_2}{d\varepsilon_v}\right)^{1/2} \tag{4}$$

$$f' = \frac{\sigma_1\left(1+1/9f^2\right)^{1/2}}{\overline{\sigma}_0} \tag{5}$$

where the three principal strains correspond to the principal strain directions $\varepsilon_1$, $\varepsilon_2$, and $\varepsilon_3$. $\varepsilon_v$ denotes the volumetric strain, while $\rho$ represents the relative density of the porous material, indicating the fraction of solid matrix present within a unit volume.

In the Shima–Oyane model, the parameters $f$ and $f'$ are critical for characterizing the yield behavior during the densification of alloy powders. The parameter $f$ can be determined through uniaxial compression tests on samples with varying levels of densification, while $f'$ can be obtained by curve fitting the relationship between the yield stress and relative density. Since the yield stress is influenced not only by the relative density but also by the temperature, the relationship between the $f'$ parameter and the temperature can be established through high-temperature uniaxial compression experiments.

### 2.2. Experiment Procedures for Model Solution and Validation

Since the yield surface in the HIP process is ellipsoidal, two independent experiments are required to determine the corresponding material parameters. In this study, uniaxial compression and interrupted HIP experiments were conducted to characterize the yield surface for the HIP simulation model, as illustrated in Figure 1. The powder used in the HIP process is a commercial-grade material with a particle size distribution ranging from 100 to 150 μm and a nearly spherical morphology, as shown in Figure 2a. The experimental samples were extracted from the preforms produced in the HIP tests described in Figure 2b and subsequently processed according to the uniaxial compression procedure. The samples were compressed by 20% and 40% at 800 °C and 1000 °C to obtain axial strain values corresponding to different densities. After compression, the density of each specimen was measured using the Archimedes drainage method, and the relationship between axial strain and density was established.

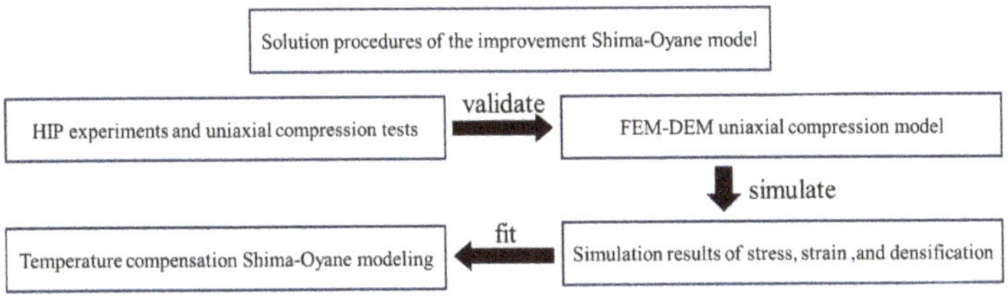


**Figure 1.** Solution procedures of parameters in the Plasticity model.

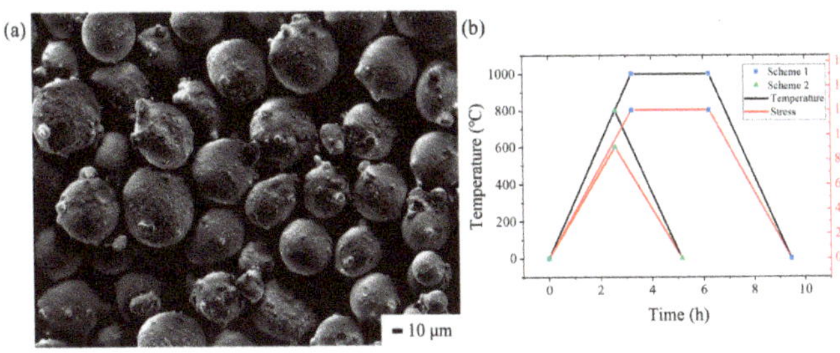


**Figure 2.** (**a**) Microscopic morphology of the powder; (**b**) HIP experimental process profile.

### 2.3. FEM-DEM Coupled Procedures for Model Solution and Prediction

The powder particle size used in HIP is on the micrometer scale. Therefore, a smaller numerical model is employed, with sufficiently small grid divisions to ensure the accuracy of the simulation results. Initially, a particle-scale finite element geometric model is constructed in the DEM. The powder material is simplified as spherical, with particle sizes ranging from 20 to 100 μm. The actual particle size distribution of the material is also considered, with the powder diameter following a normal distribution. Figure 3 shows the flow chart of the coupled thermos–mechanical mesoscopic numerical simulation of the powder HIP process, where each particle is assigned an independent mesh and treated as a distinct individual for calculation. The simulation was conducted at temperatures ranging from 700 °C to 1100 °C under an applied axial pressure of 120 MPa. The inter-particle friction coefficient was set to 0.2.

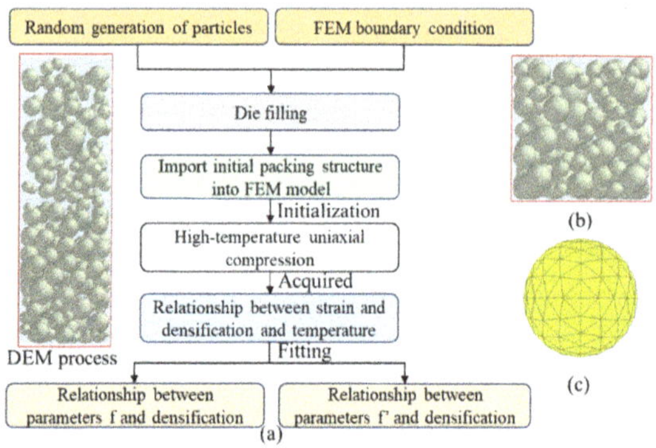


**Figure 3.** (**a**) Flow chart of mesoscopic numerical simulation of HIP process; (**b**) thermal coupling structure of particle stacking before powder HIP molding; (**c**) mesh division for individual particle.

## 3. Solution Process of Improved Powder-Yield Model

### *3.1. Validation of FEM-DEM Coupled Model*

Figure 4 shows the true stress–strain curves of a sample with a density of 0.980 at 800 °C and 1100 °C. From the figure, it can be seen that the temperature significantly impacts true stress and yield strength. From the true stress–strain data, it can be derived that when the compression temperature is 1100 °C, the yield stress of Inconel 625 alloy is 113.648 MPa and 139.570 MPa for deformation amounts of 20% and 40%, respectively. When the compression temperature is 800 °C, the yield stress of the Inconel 625 alloy is 312.925 MPa. The densification of each sample after uniaxial compression was also measured. Under compression conditions of 1100 °C and 0.01 $s^{-1}$, the densification was 0.985 after 20% compression and 0.991 after a 40% compression. Under compression conditions of 800 °C and 0.01 $s^{-1}$, the densification was 0.982 after a 20% compression.

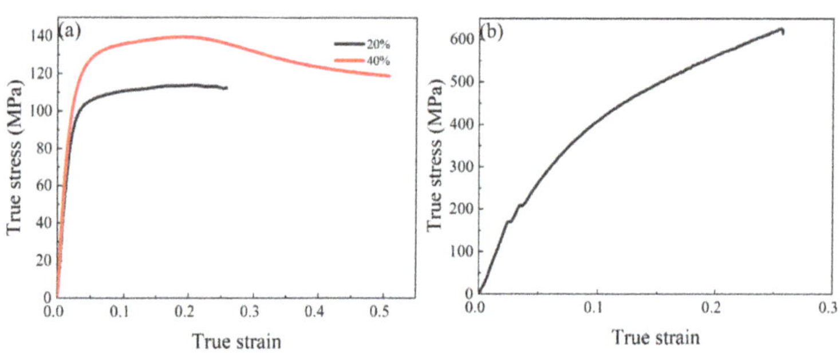


**Figure 4.** Stress–strain curves of samples with a densification of 0.980 at different temperatures: (**a**) 1100 °C; (**b**) 800 °C.

As shown in Table 1, a comparison of the strain increase corresponding to the same densification increase is presented under the same initial densification. According to Table 1, the error between experimental and simulated strain increments is within 5% for different densification increments, indicating that mesoscopic powder particle-scale simulations can effectively predict strain increments at different densification levels.

**Table 1.** Comparison of axial strain increments between simulated and experimental results at the same densification increments.

| Initial Density | Temperature | Density Increment | Strain | | Relative Error (%) |
|---|---|---|---|---|---|
| | | | Experimental Results | Simulation Results | |
| 0.980 | 1100 °C | 0.005 | 0.26 | 0.25 | 4.7 |
| 0.980 | 1100 °C | 0.013 | 0.51 | 0.53 | 3.7 |
| 0.980 | 800 °C | 0.023 | 0.26 | 0.27 | 4.3 |

Table 2 compares the yield stress obtained experimentally and through simulation at the same densification. The table shows that the yield stress obtained experimentally and the yield stress obtained through simulation differ only slightly, with an error of about 5%. This indicates that mesoscopic particle-scale simulations can effectively predict the yield strength of the Inconel 625 alloy at different densifications and temperatures.

**Table 2.** Comparison of experimental and simulated yield stresses at the same uniform density.

| Initial Density | Temperature (°C) | Yield Stress | | Relative Error (%) |
|---|---|---|---|---|
| | | Experimental Results (MPa) | Simulation Results (MPa) | |
| 0.985 | 1100 | 113.6 | 108.8 | 4.2 |
| 0.990 | 1100 | 139.6 | 132.5 | 5.1 |
| 0.982 | 800 | 625.9 | 602.0 | 3.8 |

### *3.2. Compensation of Temperature Influence on Shima–Oyane Model*

Through the validated FEM-DEM coupled high-temperature uniaxial compression model, the strain and density data of the powder at various temperatures were obtained. Figure 5a illustrates the relationship between axial strain and density during the uniaxial compression process. At lower densities, the axial strain remains close to zero, indicating that particle movement and rearrangement dominate at this stage, with minimal plastic deformation of the powder particles. As the compression progresses, the plastic deformation becomes more pronounced. Due to the softening effect at elevated temperatures, the axial strain increases more rapidly with rising temperature.

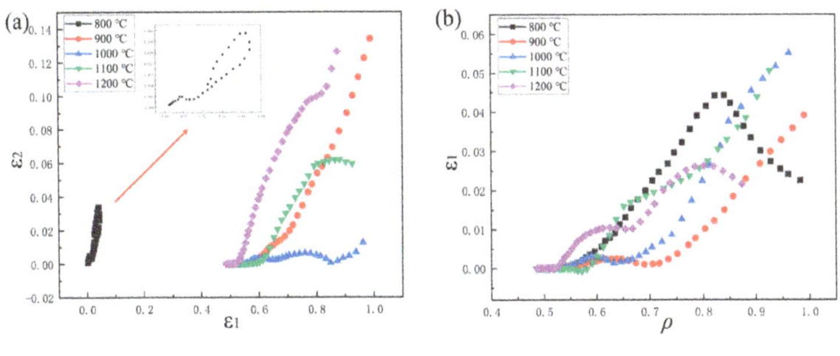


**Figure 5.** The relationships of axial strain with (**a**) the relative density and (**b**) the radial strain.

Figure 5b presents the relationship between axial strain and radial strain during uniaxial compression. At the initial stage and under low-temperature conditions, the radial strain is nearly zero, suggesting that the plastic deformation is limited and the radial expansion is suppressed. However, at higher temperatures, the plastic deformation occurs more readily due to thermal softening. As the compression continues, the radial strain at low temperatures exhibits a rapid increase, while at high temperatures, it increases gradually and smoothly. This behavior may be associated with thermally activated recovery and recrystallization processes, which are not discussed in detail in this study.

As shown in Figure 6, the value of $f$ can be calculated by substituting the experimentally verified strain and density data obtained from high-temperature uniaxial compression tests at different temperatures into Formula (4). As can be seen from the figure, the $f$ values are relatively scattered, and they overlap between different temperatures, without showing any temperature dependence.

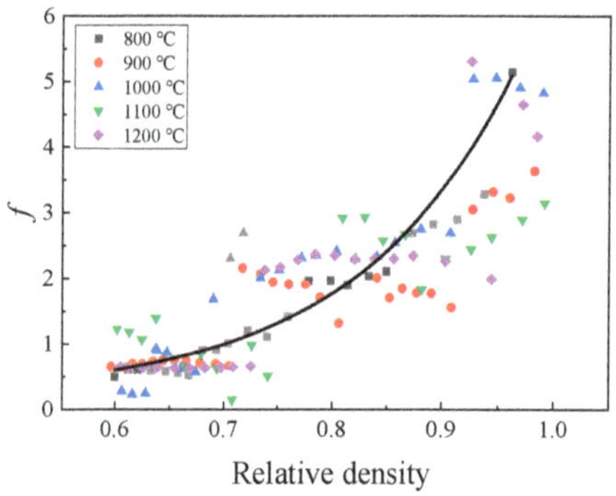


**Figure 6.** Variation in parameter $f$ with relative density.

The parameters $f'$ and $f$ are the material parameters of the powder, which are fitted through the following two formulas, which simplify both $f$ and $f'$ into functions related to density. The fitting results of parameter $f$ are shown in Figure 6.

$$f\prime = (b_1 + b_2]\rho^{b_3})^{b_4} \tag{6}$$

$$f = (q_1 + q_2\rho^{q_3})^{q_4} \tag{7}$$

where $b_1$, $b_2$, $b_3$, and $b_4$ are the material constants related to temperature; $q_1$, $q_2$, $q_3$, and $q_4$ are the material constants. The fitting result is as follows:

$$f = \left(0.91632 + 0.17049\,]\rho^{1.99964}\right)^{22.79806} \tag{8}$$

The yield strength of the fully dense Inconel 625 alloy matrix is 500 MPa. Combining the values of $\sigma_1$ at different densities and temperatures, the corresponding values $f$ for different densifications are calculated and the parameters $f'$ can be determined by substituting into Equation (8), as shown in Figure 7. Since the parameter $f'$ is related to stress and stress is affected by temperature, for powders with the same densification, higher temperature results in lower stress. Thus, the parameter $f'$ also decreases. From the figure, it can also be seen that at different temperatures, the parameter $f'$ increases with densification, and as densification increases, the rate of increase accelerates. Therefore, the calculated parameter $f'$ is fitted separately for each temperature, with the fitting results shown in Equation (9).

$$\begin{cases} f'_{T=800} = \left(0.99801 + 0.00552\,]\rho^{9.07827}\right)^{104.74863} \\ f'_{T=900} = \left(0.98668 + 0.01521\,]\rho^{4.4382}\right)^{51.48709} \\ f'_{T=1000} = \left(0.96541 + 0.02242\,]\rho^{3.83392}\right)^{28.93245} \\ f'_{T=1100} = \left(0.98574 + 0.00858\,]\rho^{6.74518}\right)^{86.32917} \\ f'_{T=1200} = \left(0.93754 + 0.02054\,]\rho^{3.97408}\right)^{31.99983} \end{cases} \quad (9)$$

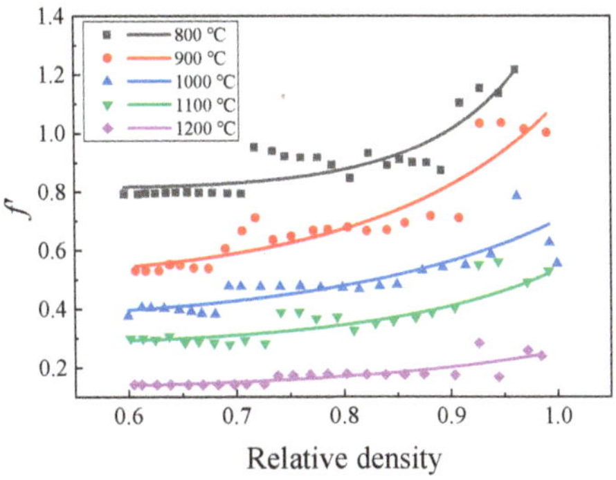


**Figure 7.** Variation in parameter $f'$ with relative density.

The most suitable temperature-dependent expression functions for the four material parameters are found, and the material constants of the Inconel 625 alloy are expressed as temperature-dependent functions.

$$\begin{cases} b_1 = C_1 + C_2T^2 + C_3T^4 + C_4T^6 \\ b_2^{-1} = D_1 + D_2T + D_3T^2 + D_4T^3 \\ b_3 = E_1 + E_2T^2 \ln T + E_3T^{2.5} + E_4T^3 \\ b_4^2 = G_1 + G_2T + G_3T^2 + G_4T^3 \end{cases} \quad (10)$$

The variation trends of the material constants $b_1$, $b_2$, $b_3$, and $b_4$ after the function fitting are shown in Figure 8.

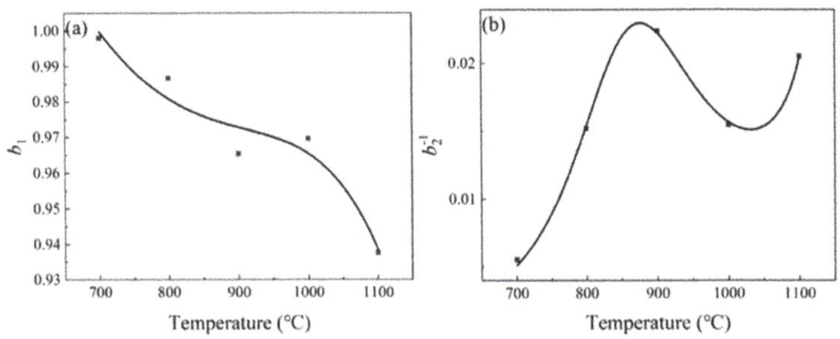


**Figure 8.** *Cont.*

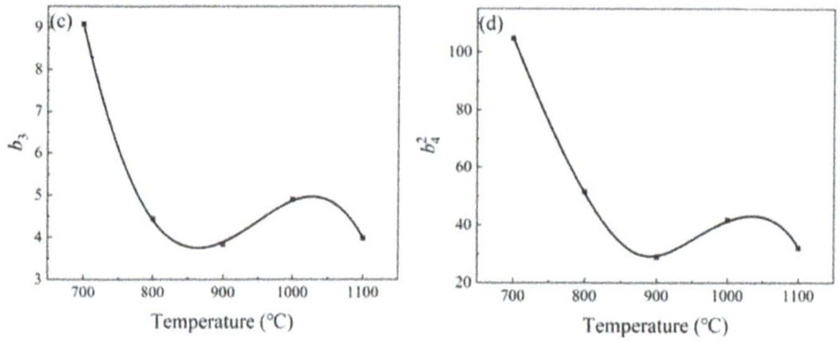


**Figure 8.** Fitting of material constants (**a**) $b_1$, (**b**) $b_2$, (**c**) $b_3$, and (**d**) $b_4$ to temperature.

Based on the above formulas, the coefficients are determined through nonlinear function fitting using the Origin 2019b software, as shown in Table 3.

**Table 3.** Correlation coefficients of Inconel 625 alloy $b_1$, $b_2$, $b_3$, and $b_4$ fitting functions.

| | $b_1$ | | $b_2$ | | $b_3$ | | $b_4$ |
|---|---|---|---|---|---|---|---|
| $C_1$ | 1.25004 | $D_1$ | 10,295.841 | $E_1$ | 164.52288 | $G_1$ | 601,825.6 |
| $C_2$ | $-9.42466 \times 10^{-7}$ | $D_2$ | −32.66279 | $E_2$ | −0.00721 | $G_2$ | −1891.7101 |
| $C_3$ | $1.09527 \times 10^{-12}$ | $D_3$ | 0.03449 | $E_3$ | 0.00024 | $G_3$ | 1.97538 |
| $C_4$ | $-4.37362 \times 10^{-19}$ | $D_4$ | $-1.20587 \times 10^{-5}$ | $E_4$ | $-2.7258 \times 10^{-6}$ | $G_4$ | −0.00068 |

### 3.3. Construction of Stress–Strain–Temperature–Density Relationships

Based on the obtained parameters at various densities and temperatures, a yield stress map for the nickel-based alloy was constructed, as shown in Figure 9. In this figure, the contour lines represent variations in the yield strength with the temperature at a given density, and the labeled values indicate the corresponding yield stresses. The color gradient visually illustrates the yield stress distribution. As observed, the yield stress decreases with the increasing temperature, reaching a minimum of 52.087 MPa at 1200 °C. Conversely, as the density increases, the yield strength of the material also rises. Thus, the yield strength is jointly influenced by both the temperature and the density, exhibiting a nonlinear relationship. Although a higher density contributes to an increase in the yield stress, the dominant factor influencing its variation is the reduction caused by elevated temperatures. At a constant temperature, the maximum variation in the yield stress is 101.214 MPa, while at a constant density, the maximum variation is 244.515 MPa.

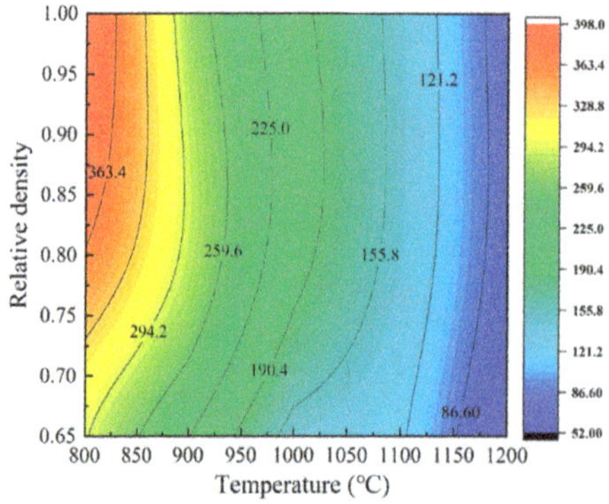


**Figure 9.** Yield strength as a function of temperature and densification.

## 4. Predication of Powder Densification Behaviors and Validation Evaluation

### 4.1. Presentation of Stress–Strain–Temperature–Density Relationships by Solved Model

#### 4.1.1. Macroscopic Analysis of the Experimental Results

During the powder HIP process, powder densification is mainly driven by the deformation of the casing. Uniform deformation of the casing results in a uniformly densified final part. However, suppose the local deformation in the casing is too large or too small; in that case, it can not only lead to a casing rupture but also cause localized densification and non-uniformity in the formed part. Therefore, the deformation of the casing directly affects the total performance of the final part. The green bodies obtained from the two different HIP processes shown in Figure 2 are illustrated in Figure 10. The figure shows that after holding at 1000 °C and 120 MPa for 180 min under process scheme 1, the shrinkage deformation of the casing in the cylindrical part is noticeably greater than that of the cylindrical part formed with process scheme 2. To quantitatively compare the casing shrinkage deformation under the two process schemes, axial and radial measurements of the casing were performed. After measurements, it was found that the casing processed under process scheme 1 experienced an axial shrinkage of 13.24 mm and a radial shrinkage of 12.50 mm. After processing with process scheme 2, the casing experienced an axial shrinkage of 5.32 mm and a radial shrinkage of 4.46 mm. It is evident that the process parameters significantly affect the casing shrinkage and play a decisive role in the powder densification process.

**Figure 10.** Comparison of two post-HIP program capsule sets.

During the HIP process, powder densification is primarily driven by the deformation of the encapsulating structure. The degree of densification within the powder compact closely correlates with the extent of capsule deformation. Figure 11 provides a comparative visualization of the central axial cross-sections obtained from experimental observation and numerical simulation. As illustrated, the deformation profiles of the truncated capsule in the central axial direction are highly consistent, indicating good agreement between the model and experimental results. To further evaluate the fidelity of the simulation, two characteristic lines were selected: L1, extending diagonally from the powder corner to the center point, and L2, corresponding to the axis of symmetry through the center. Three feature nodes were designated along each line—A1, A2, and A3 on L1, and B4 and B5 on L2—with A3 located precisely at the intersection of L1 and L2, at the powder's geometric center. Samples with dimensions of ø10 mm × 12 mm were extracted from each feature location. The sample densities were determined via the Archimedes drainage method, with each mass measurement repeated multiple times and averaged to enhance the measurement reliability. The measured densities were quantitatively compared with the simulated results at the corresponding nodes. The close match between the experimental and simulated densities validates the predictive capability of the simulation model.

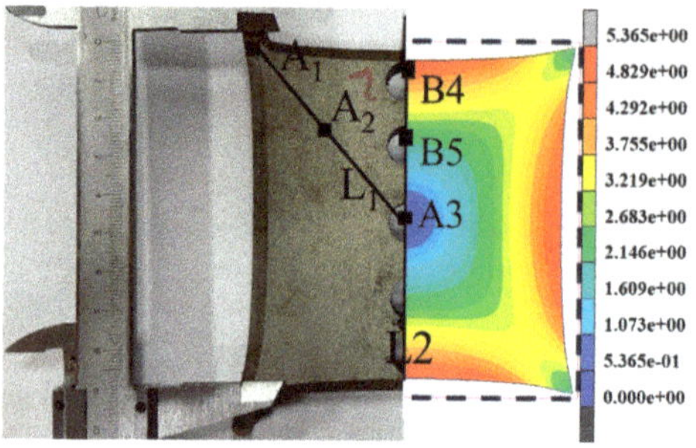


**Figure 11.** Macro-deformation comparison.

#### 4.1.2. Evolution of the Relative Density

Figure 12 illustrates the density evolution of five feature nodes during the HIP process, which consists of three stages: heating and pressurization, holding and pressure maintenance, and cooling and depressurization. In the initial stage of the HIP, the density at each node decreases due to the thermal expansion of the powder, while the applied pressure is insufficient to counteract this expansion, resulting in a temporary increase in volume. However, this stage is brief, and the reduction in density is minimal compared to the overall increase. As the HIP process progresses, the temperature and pressure rise. At 7000 s, the density at each node begins to increase rapidly. The applied pressure becomes sufficient to exceed the powder's yield strength, causing shrinkage and deformation. Since the initial powder packing is relatively loose, a significant plastic deformation occurs, leading to a more compact internal arrangement. For feature line L1, nodes A1, A2, and A3 exhibit similar density growth patterns, with densities steadily increasing after the initial HIP stage. The final densities of A2 and A3 reach approximately 0.97, higher than A1, due to A2 and A3 being closer to the center of the powder, while A1 is affected by its position at the edge, near the corner and end caps, resulting in a lower density. Nodes B4, B5, and A3 along line L2 show similar trends, with their density curves largely coinciding. After an initial reduction, the density increases steadily, but the rate of increase slows during the later holding stage as densification approaches saturation. At this point, the densification is primarily driven by diffusion and creep, and the final density reaches approximately 0.97.

#### 4.1.3. Microstructure of HIP Sample

As shown in Figure 13, the microstructure of the billet, sampled from the center cross-section at a temperature of 1100 °C, a pressure of 120 MPa, and a holding time of 3 h, exhibits a density of 0.980. In comparison, Figure 12 shows that the simulated density at the center of the cylindrical part reaches 0.972. The minimal deviation between the simulated and experimental values confirms the accuracy of the simulation results. Figure 13 visually presents the microstructural features, showing a markedly more compact arrangement of powder particles. The pores are nearly eliminated, and the particles exhibit substantial plastic deformation, indicating effective densification during the HIP process. The billet was processed by HIP, a technique that applies high temperature and isostatic gas pressure to densify materials and eliminate internal pores. After HIP, the density of the powder material has substantially improved, with some regions showing almost complete elimination of porosity. However, original particle boundaries remain visible, and distinct grain boundary features can still be observed, as shown in the yellow circle. This indicates that, despite the applied process parameters, some powder particles did not fully fuse during HIP. The causes of this incomplete fusion are complex and are beyond the scope of this paper.

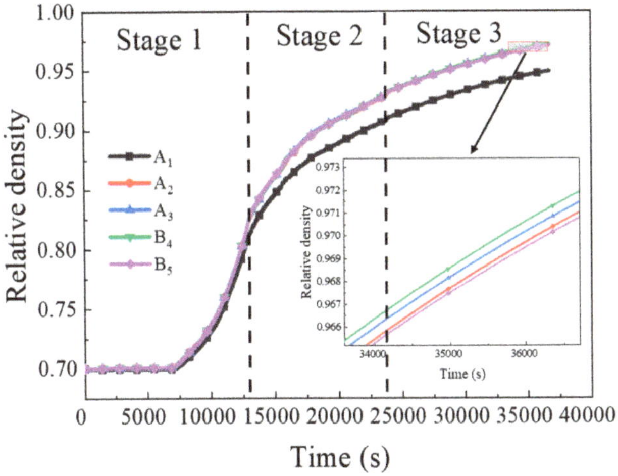


**Figure 12.** Evolution of the relative density of different feature points over time.

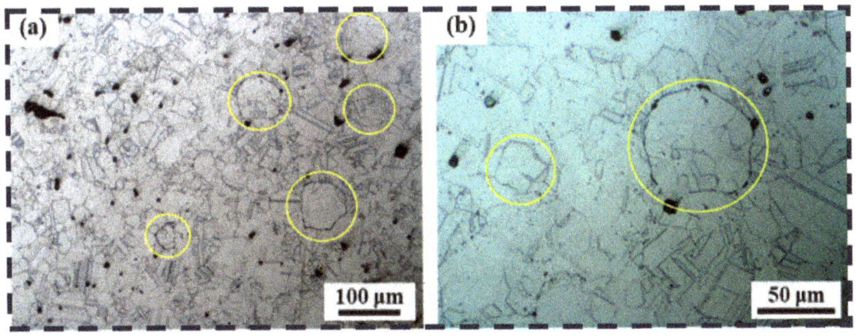


**Figure 13.** Microstructure of HIP parts: (**a**) 100X and (**b**) 200X.

*4.2. Validation Evaluation of Improved Powder-Yield Model*

The results comparing the densities from the experimental and numerical simulations are shown in Figure 14. It can be observed that, under the same parameter conditions, the densities at each feature location of the cylinders formed by HIP are similar to the results of the numerical simulations, with the densities of the five feature nodes all exceeding 0.95. Specifically, due to the influence of the capsule's edges and corners, the experimentally obtained cylinder has the lowest density of 0.96 at feature node A1, while the densities at the other four feature nodes are around 0.98. The density variation between the nodes is no more than 0.02, indicating good homogeneity of the HIPed powder. As shown in Figure 14, the relative errors between the experimental and simulation results for the five feature points are 1.07%, 1.15%, 1.35%, 1.09%, and 1.21%, respectively, all within 2%. Overall, the simulated densities are slightly lower than the experimental densities, which can be attributed to the nature of the macroscopic simulation, where the powder is treated as a continuous porous medium. This differs from the discrete particle nature of the HIP process, which ignores the effects of powder particle rearrangement during the initial stage and the diffusion creep at high temperatures during the final stage. Nevertheless, the comparison of the numerical simulation and experimental results for the macroscopic deformation and densification at feature locations shows that the numerical simulation method used in this study provides an accurate prediction of the densification behavior of powders during powder HIP.

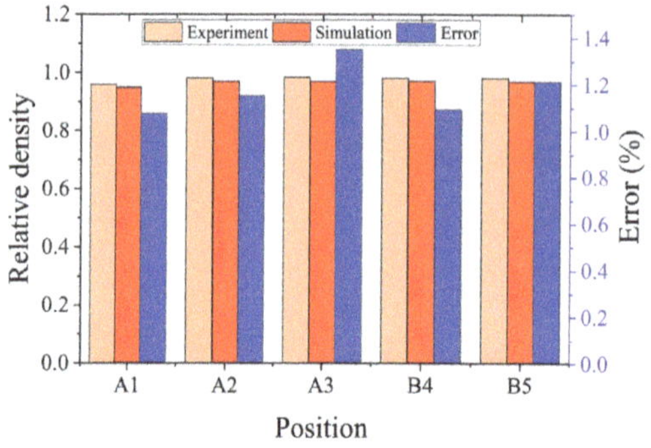


**Figure 14.** Comparison of simulation results with experimental results.

## 5. Conclusions

Based on the DEM theory, a mesoscopic particle-scale numerical model of the powder HIP was constructed. Based on the numerical model, the stress and strain parameters of the particles in the forming process under different temperature and pressure conditions were extracted and characterized as a function of the degree of densification. Uniaxial compression experiments verified the above parameters' HIPed samples at different temperatures. Then, a temperature-dependent Shima–Oyane model based on the validation parameters was developed. Key findings were as follows:

The parameters relating to the stress, strain, and densification at different temperatures and pressures were obtained based on the mesoscopic particle-scale model. The strain increments and stress at different densification levels were compared by conducting uniaxial compression experiments on samples at different temperatures. The relative error between the experimental and simulation results was within 5.1%. The mesoscopic particle-scale simulations accurately determined the parameters related to the stress, strain, and densification.

The parameters relating to the stress, strain increments, and densification at different temperatures (800–1100 °C) were obtained based on the mesoscopic particle-scale model. A temperature-compensated Shima–Oyane model was developed. Based on the parameters $f$ and $f'$, a three-dimensional relationship graph illustrating the correlation between the yield stress, density, and temperature was constructed. To validate the proposed temperature-compensated Shima–Oyane model, the deformation behavior and density distribution of the powder components formed via the HIP were investigated. The results showed that the simulated final deformation closely matched the experimental observations, and the relative errors in the predicted densities across different regions were within 2%. These findings confirm that the constructed macroscopic finite-element model can reliably predict the densification behavior during the HIP process.

Although this study presents a mesoscopic–macroscopic coupled analysis of the HIP forming process under various parameter conditions, it should be acknowledged that the Shima–Oyane model is limited in capturing the fundamental densification mechanisms of hot isostatic pressing, such as diffusional creep and grain boundary diffusion. Physically based creep models are better suited to represent the microstructural deformation mechanisms that dominate during the HIP process. Additionally, the current DEM model provides a foundation for particle-scale studies of the densification process during hot isostatic pressing. These aspects will be the primary focus of the authors' future work.

**Author Contributions:** Conceptualization, G.Q. and W.R.; Methodology, G.Q. and W.R.; Validation, Q.J., Y.Y. and Y.Z.; Formal Analysis, W.D.; Investigation, Y.Z.; Resources, G.Q.; Data Curation, W.R., W.D., Q.J. and Y.Y.; Writing—Original Draft Preparation, W.R.; Writing—Review & Editing, G.Q. and W.R.; Visualization, W.R. and W.D.; Supervision, G.Q.; Funding Acquisition, G.Q. All authors have read and agreed to the published version of the manuscript.

**Funding:** The authors are grateful for the supports from National Natural Science Foundation of China (52175287) and Chongqing Natural Science Foundation General Project (CSTB2022NSCQ-MSX0593).

**Data Availability Statement:** The original contributions presented in this study are included in the article. Further inquiries can be directed to the corresponding author.

**Conflicts of Interest:** Authors Weiwei Dai and Qian Jiang were employed by COSCO Shipping Marine Equipment & Spares (Nanjing) Co., Ltd. Author Guozheng Quan was a part-time employee of COSCO Shipping Marine Equipment & Spares (Nanjing) Co., Ltd. The remaining authors declare that the research was conducted in the absence of any commercial or financial relationships that could be construed as a potential conflict of interest.

## References

1. Boris, B.I.; Genrikh, G.S.; Oleg, F.C. Prospect of HIP application in aerospace and power plant industry. In Proceedings of the International Conference on Hot Isostatic Pressing, Andover, MA, USA, 20–22 May 1996.
2. Atkinson, H.V.; Davies, S. Fundamental aspects of hot isostatic pressing: An overview. *Metall. Mater. Trans. A* **2000**, *31*, 2981–3000. [CrossRef]
3. Zhang, H.; Wang, A.; Wen, Z.; Yue, Z.; Zhang, C. Effects of Hot Isostatic Pressing (HIP) on Microstructure and Mechanical Properties of K403 Nickel-Based Superalloy. *High Temp. Mater. Process.* **2016**, *35*, 463–471. [CrossRef]
4. ElRakayby, H.; Kim, K. Effect of glass container encapsulation on deformation and densification behavior of metal powders during hot isostatic pressing. *Int. J. Mater. Form.* **2017**, *11*, 517–525. [CrossRef]
5. Semiatin, S.L. An Overview of the Thermomechanical Processing of $\alpha/\beta$ Titanium Alloys: Current Status and Future Research Opportunities. *Metall. Mater. Trans. A* **2020**, *51*, 2593–2625. [CrossRef]
6. Kim, H.S. Densification mechanisms during hot isostatic pressing of stainless steel powder compacts. *J. Mater. Process. Technol.* **2002**, *123*, 319–322. [CrossRef]
7. Xue, Y.; Lang, L.; Bu, G.; Li, L. Densification Modeling of Titanium Alloy Powder during Hot Isostatic Pressing. *Sci. Sinter.* **2011**, *43*, 247–260. [CrossRef]
8. Abdelhafeez, A.; Essa, K. Influences of Powder Compaction Constitutive Models on the Finite Element Simulation of Hot Isostatic Pressing. *Procedia CIRP* **2016**, *55*, 188–193. [CrossRef]
9. Wei, Q.S.; Xue, P.J.; Liu, G.C.; Lu, H.; Huang, J.; Shi, Y.S. Simulation and verification of near-net shaping a complex-shaped turbine disc by hot isostatic pressing process. *Int. J. Adv. Manuf. Technol.* **2014**, *74*, 1667–1677. [CrossRef]
10. Zhou, S.; Song, B.; Xue, P.; Cai, C.; Liu, J.; Shi, Y. Numerical simulation and experimental investigation on densification, shape deformation, and stress distribution of Ti6Al4V compacts during hot isostatic pressing. *Int. J. Adv. Manuf. Technol.* **2017**, *88*, 19–31. [CrossRef]
11. Bu, G.L.; Lang, L.H.; Wang, G.; Song, Y.; Xu, Q.Y. Forming and Simulation of Titanium Alloy Ti-6Al-4V by Hot Isostatic Pressing. *Adv. Mater. Res.* **2013**, *848*, 50–54. [CrossRef]
12. Wang, M.; Yin, Y.J.; Zhou, J.X.; Ji, X.Y.; Nan, H.; Xue, P.J.; Shi, Y.S. A multi-scale study of Inconel 625 powders HIP process and construction of HIP maps. *Int. J. Adv. Manuf. Technol.* **2017**, *90*, 3055–3066. [CrossRef]
13. Liu, G.C.; Shi, Y.S.; Wei, Q.S.; Wang, J.W. Finite Element Analysis of Pressure Influence on Densification of Titanium Alloy Powder under Hot Isostatic Pressing. *Key Eng. Mater.* **2010**, *450*, 206–209. [CrossRef]
14. Geindreau, C.; Bouvard, D.; Doremus, P. Constitutive behaviour of metal powder during hot forming. Part I: Experimental investigation with lead powder as a simulation material. *Eur. J. Mech. A Solids* **1999**, *18*, 581–596. [CrossRef]
15. Van Nguyen, C.; Bezold, A.; Broeckmann, C. Anisotropic shrinkage during hip of encapsulated powder. *J. Mater. Process. Technol.* **2015**, *226*, 134–145. [CrossRef]

16. Han, P.; An, X.Z.; Wang, D.F.; Fu, H.T.; Yang, X.H.; Zhang, H.; Zou, Z.S. MPFEM simulation of compaction densification behavior of Fe-Al composite powders with different size ratios. *J. Alloys Compd.* **2018**, *741*, 473–481. [CrossRef]
17. Wang, B.; Pan, K.J.; Gao, S.; Wu, S.X.; Zhao, C.; Luo, X.; Peng, Q.; Sun, M.H.; Li, D.D.; Li, N.; et al. Synergistic densification mechanism of irregular Ti powder during CIP: 3D MPFEM simulation with real-shape particles. *Mater. Des.* **2024**, *246*, 113368. [CrossRef]
18. Xu, L.; Wang, Y.; Li, C.; Ji, G.; Mi, G. MPFEM simulation on hot-pressing densification process of SiC particle/6061Al composite powders. *J. Phys. Chem. Solids* **2021**, *159*, 110259. [CrossRef]
19. Zhang, W.; Yuan, C.; Xiao, W.; Gong, X.; Hai, B.; Chen, R.; Zhou, J. MPFEM investigation on densification and mechanical structures during ferrous powder compaction. *Adv. Powder Technol.* **2024**, *35*, 104700. [CrossRef]
20. Zou, Y.; An, X.; Zou, R. Investigation of densification behavior of tungsten powders during hot isostatic pressing with a 3D multi-particle FEM approach. *Powder Technol.* **2020**, *361*, 297–305. [CrossRef]
21. Harthong, B.; Imbault, D.; Dorémus, P. The study of relations between loading history and yield surfaces in powder materials using discrete finite element simulations. *J. Mech. Phys. Solids* **2012**, *60*, 784–801. [CrossRef]
22. Loidolt, P.; Ulz, M.H.; Khinast, J. Modeling yield properties of compacted powder using a multi-particle finite element model with cohesive contacts. *Powder Technol.* **2018**, *336*, 426–440. [CrossRef]
23. Güner, F.; Sofuoglu, H.; Cora, Ö.N. An investigation of contact interactions in powder compaction process through variable friction models. *Tribol. Int.* **2016**, *96*, 1–10. [CrossRef]
24. Xin, X.J.; Jayaraman, P.; Jiang, G.; Wagoner, R.H.; Daehn, G.S. Explicit finite element method simulation of consolidation of monolithic and composite powders. *Met. Mater. Trans. A* **2002**, *33*, 2649–2658. [CrossRef]
25. Zhou, J.; Xu, H.; Zhu, C.; Wang, B.; Liu, K. Comparisons Between 2D and 3D MPFEM Simulations in Modeling Uniaxial High Velocity Compaction Behaviors of Ti-6Al-4V Powder. *Arch. Metall. Mater.* **2022**, *67*, 57–65. [CrossRef]
26. Zhou, J.; Zhu, C.; Zhang, W.; Ai, W.; Zhang, X.; Liu, K. Experimental and 3D MPFEM simulation study on the green density of Ti-6Al-4V powder compact during uniaxial high velocity compaction. *J. Alloys Compd.* **2020**, *817*, 153226. [CrossRef]
27. Elguezabal, B.; Martínez-Esnaola, J.; Soler, R.; Paños, E.; Alkorta, J. A multiscale material model for metallic powder compaction during hot isostatic pressing. *Powder Technol.* **2023**, *425*, 118599. [CrossRef]
28. Li, M.; Lim, C.V.S.; Zou, R.; An, X.; Wang, D. Multi-particle FEM modeling on hot isostatic pressing of Ti6Al4V powders. *Int. J. Mech. Sci.* **2021**, *196*, 106288. [CrossRef]
29. Li, M.; Jia, Q.; Li, C.; Guo, Q.; An, X.; Zou, Q.; Zhou, X.; Fu, H.; Zhang, H.; Yang, X.; et al. 3D particulate-scale numerical investigation on hot isostatic pressing of W-Cu composites. *Powder Technol.* **2023**, *415*, 118150. [CrossRef]
30. Oyane, M.; Shima, S.; Kono, Y. Thoery of Plasticity for Porous Metals. *Bull. JSME* **1973**, *16*, 1254–1262. [CrossRef]
31. Shima, S.; Oyane, M. Plasticity theory for porous metals. *Int. J. Mech. Sci.* **1976**, *18*, 285–291. [CrossRef]

*metals*

MDPI

*Article*

# Experimental Investigation of the Interplay Between Al-, B-, and Ti-Nitrides in Microalloyed Steel and Thermodynamic Analysis

**Markus Führer [1,*], Sabine Zamberger [2], Christoph Seubert [1] and Erwin Povoden-Karadeniz [1,3]**

1 Christian Doppler Laboratory for Interfaces and Precipitation Engineering CDL-IPE, Institute of Materials Science and Technology, TU Wien, Getreidemarkt 9, 1060 Vienna, Austria; christoph.seubert@tuwien.ac.at (C.S.); erwin.povoden-karadeniz@tuwien.ac.at (E.P.-K.)

2 Voestalpine Forschungsservicegesellschaft Donawitz GmbH, Kerpelystraße 199, 8700 Leoben, Austria; sabine.zamberger@voestalpine.com

3 Institute of Materials Science and Technology, TU Wien, Getreidemarkt 9, 1060 Vienna, Austria

* Correspondence: markus.fuehrer@tuwien.ac.at

**Abstract:** Aluminum, boron, and titanium microalloyed into high-strength low-alloy boron steel exhibit a complex interplay, competing for nitrogen, with titanium demonstrating the highest affinity, followed by boron and aluminum. This competition affects the formation and distribution of nitrides, impacting the microstructure and mechanical properties of the steel. Titanium protects boron from forming BN and facilitates the nucleation of acicular ferrite, enhancing toughness. The segregation of boron to grain boundaries, rather than its precipitation as boron nitride, promotes the formation of martensite and thus the through-hardenability. Aluminum nitride is critical in controlling grain size through a pronounced pinning effect. In this study, we employ energy- and wavelength-dispersive X-ray spectroscopy and computer-aided particle analysis to analyze the phase content of 12 high-purity vacuum induction-melted samples. The primary objective of this study is to correctly describe the microstructural evolution in the Fe-Al-B-Ti-C-N system using the Calphad approach, with special emphasis on correctly predicting the dissolution temperatures of nitrides. A multicomponent database is constructed through the incorporation of available binary and ternary descriptions, employing the Calphad approach. The experimental findings regarding the solvus temperature of the involved nitrides are employed to validate the accuracy of the thermodynamic database. The findings offer a comprehensive understanding of the relative phase stabilities and the associated interplay among the involved elements Al, B, and Ti in the Fe-rich corner of the system. The type and size distribution of the stable nitrides in microalloyed steel have been demonstrated to exert a substantial influence on the properties of the material, thereby rendering accurate predictions of phase stabilities of considerable relevance.

**Keywords:** HSLA; boron-steel; thermodynamics

## 1. Introduction

High-strength low-alloy (HSLA) boron steels are widely recognized for their enhanced mechanical properties, which are achieved through the precise control of their microstructure [1]. Reaching the full potential of alloying elements in microalloyed steel is dependent on correctly understanding the variables processing, thermodynamics, and kinetics [2] to reach the desired mechanical properties with reasonable costs [1]. The addition of boron to microalloyed steel has been shown to enhance through-hardenability. Furthermore, the formation of bainite and martensite is promoted by alloying the steel with only about ten wt.-ppm of boron [3–5]. The phenomenon referred to as the 'boron effect' has been observed to enhance through-hardenability rather than hardness [6]. The underlying mechanism of this advantageous effect involves the retention of boron in solid solution in the matrix, thereby preventing the formation of boron nitride (BN) [7]. In order to protect the steel from the occurrence of this second phase formation, i.e., binding the nitrogen instead of BN, aluminum or titanium are utilized [7]. Titanium, with its high affinity for nitrogen [8], primarily forms titanium nitrides (TiN), which are known to refine the grain structure and improve toughness [9]. Aluminum nitride (AlN) has been shown to be particularly effective in controlling grain size due to its pronounced pinning effect, which is known to inhibit grain growth [10]. The interplay between these elements and their respective roles in the nitride stabilization is critical in determining the final microstructure and mechanical properties of the steel. The interactive role of Al, B, and Ti in the formation and dissolution of nitride precipitates, as well as the associated content of boron in solid solution for the consolidation of martensite, is a subject of considerable concern [7].

In their study, Wang et al. [8] examined the precipitation of AlN, BN, and TiN in boron-bearing steel by using available solubility product descriptions from the literature for describing the stability of the nitrides, concluding that the precipitation of TiN occurs predominantly before the precipitation of BN and AlN. In a related study, Gontijo et al. [11] report on the effects of combined addition of B and Ti in microalloyed steel. The authors find that the formation of TiN is preferential over BN and AlN with regard to ductility behavior. This phenomenon occurs due to segregated B at grain boundaries, which retards the formation of ferrite films at the austenite grain boundaries. To date, a detailed investigation of the competing stabilities of the nitrides from Al, B, and Ti in microalloyed steel, either experimental or simulated, is lacking.

The objective of this study is to apply Calphad [12] in order to elucidate and validate the thermodynamic behavior and microstructural evolution of microalloyed steels containing aluminum, boron, and titanium nitrides in the Fe-rich Fe-Al-B-Ti-C-N system. Experimentally, a combination of advanced analytical techniques, including energy-dispersive X-ray spectroscopy (EDS) in combination with computer-aided EDS and wavelength-dispersive X-ray spectroscopy, is employed to analyze 12 high-purity vacuum induction-melted samples. Furthermore, the material is subjected to annealing under equilibrium-like durations in a dilatometer at varying temperatures. This procedure is employed to ascertain the solvus temperatures of the implicated precipitates. The experimental results are used to validate the accuracy of the integration of binary and ternary reassessments for multi-component microalloyed steel [13–17] in the thermodynamic multi-component database ME-Fe_CDLIPE2.0.2.tdb [18]. Particular emphasis is put on the solvus temperatures (Tsol) of the relevant nitrides as a function of varying chemical compositions in the Fe-rich side in diluted systems. We provide an assessed Gibbs energy description of the BN phase, giving high weight to an accurate solubility description.

The alloying of Al, B, and Ti influences the properties of the material and determines the type and distribution of the stable nitrides. These nitrides subsequently influence the characteristics of the material. The utilization of Al for deoxidation in steel was predominant until the discovery regarding the pivotal influence of AlN precipitates on steel properties. The ability of AlN to impede austenite grain coarsening through the pinning of grain boundaries [19] is well-documented. Conversely, the presence of excessive primary AlN has been shown to induce embrittlement [10,20]. The stoichiometric phase [21] AlN possesses a hexagonal wurtzite structure with a space group of P63mc. Upon cooling, AlN begins to become stable in the austenite ($\gamma$) region of microalloyed steel, predominantly nucleating heterogeneously at grain boundaries [10] due to a high volumetric misfit of 70% between the $\gamma$-matrix and AlN-precipitate [10,22,23]. In the ferrite ($\alpha$) phase, the thermomechanical processing route and the nominal Al and N contents [24–27] are pivotal factors in determining the initial phase, representing metastable cubic rock-salt structured AlN, with a size ranging from 5 to 10 nanometers. This metastable AlN can subsequently be converted into the stable hexagonal AlN modification [28].

The "boron effect" has been demonstrated to be beneficial; however, it only occurs when solute B atoms are present at grain boundaries, which makes the formation of BN a matter of significant concern. BN occurs as hexagonal BN type (Pearson symbol hP4 [29]) in steel [7] and the nucleation of BN has been reported to occur at austenite grain boundaries [30]. The size of the BN precipitates, ranging from 5 to 30 μm, is reported to be dependent on the time–temperature profile of the applied heat treatments [31]. However, this size range is too coarse to have any beneficial effect on the properties of the material [7].

Titanium exhibits a strong affinity for carbon and nitrogen, which leads to the formation of TiN and TiC precipitates. TiN nucleates as face-centered cubic (fcc) phase, often referred to as MX phase, with M being Ti, V, or Nb and X being C and N [2]. The high thermal and chemical stability of the carbonitride results in a high Tsol of the precipitates, even at low fractions of Ti [32]. Carbonitrides have been shown to impede grain growth and recrystallization, which in turn alters the transformation characteristics and improves the properties of the material [9]. The beneficial influence of Ti precipitates is only possible in fully killed steels that have undergone deoxidation with Al due to the strong deoxidizing characteristics and the subsequent formation of titanium oxide [33].

Abnormal grain growth (AGG) in steel is a phenomenon in which certain grains grow significantly larger than others during heat treatment processes such as annealing [34]. This can result in a bimodal grain size distribution that affects the properties of the material. AGG typically occurs in steel at elevated temperatures after the dissolution of precipitates such as Nb or V carbonitrides, which are known to exert a strong solute drag effect on grain boundary mobility [35–37].

## 2. The Thermodynamics of the Fe-C-N-A-B-Ti System in the Fe-Rich Corner in Microalloyed Steel

Typically, the available binary and ternary phase diagrams of a multi-component system encompass the broad spectrum of compositions well, though they do not concentrate on a specific composition space corner. In the case of microalloyed steel with diluted fractions of certain elements, this can result in substantial discrepancies between the predicted and observed phase stability [38]. It is therefore imperative to accurately delineate the thermodynamics and phase boundaries of the constituent phases within the dilute microalloying range to ensure the efficacy of thermokinetic simulations with high predictive capabilities. This study employs available thermodynamic descriptions of binary and ternary systems, combining them into a comprehensive thermodynamic database. The objective of the database setup is to accurately predict the phase stabilities for the Fe-rich side of the multicomponent system Fe-C-N-Al-B-Ti, with a particular focus on the interaction of Al, B, and Ti with nitrogen. This particularly evaluates whether the approach of combining binary and ternary thermodynamics, while keeping further extension to the quaternary and quinary subsystems and beyond ideal (i.e., without additional multicomponent excess Gibbs energies optimizations) yields sufficient predictive power for technological alloys, which would be represented by low deviations between experimental data and calculated results.

### 2.1. Calphad Modeling

For the establishment of the thermodynamic database, the CALPHAD (CALculation of PHAse Diagrams) approach [12,39] is used. In multi-component systems, it is essential to characterize the thermodynamics of all lower-order systems accurately. For pure components, the thermodynamic data is taken from Dinsdale [40]. The Redlich–Kister–Muggianu model for the Gibbs excess energy description for the phases liquid (L), bcc_A2 (bcc, ferrite, $\alpha$-Fe), and fcc_A1 (fcc, austenite, $\gamma$-Fe) is used. The liquid phase uses a random substitutional solution model. Phases bcc and fcc are described by a regular solution model. Carbides and nitrides are described with sublattice models. The following section provides a critical review of the composing subsystems.

### 2.2. Subsystems with Relevance in Microalloyed Steel

Fe-based databases are determined by the binary systems containing Fe, i.e., Fe-C, Fe-N, Fe-B, Fe-Ti, and Fe-Al. Intrinsic stabilities of non-ferrous nitrides and carbo-nitrides are relevant, so the Al-N, Al-C, Al-Ti, Ti-C, Ti-N, and B-N systems are added. The Fe-C system, actually representing the steel matrix for the precipitation, is of significant importance to the iron and steel industry, as evidenced by the extensive reviews conducted [13,41,42]. The assessment by Gustafson [13] has gained considerable recognition within Calphad databases [43]. In addition to the conventional liquid, bcc, and fcc phases, several iron carbides have been identified, with Fe3C (cementite) [13,44] being the sole metastable carbide. In fact, competitive nitrogen solubility in iron, versus nitride stabilization, is relevant, and thus also the Fe-N system, with the description by Frisk proving reliable [14]. A similar question concerns boron solubility in the steel matrix, versus boride stabilization, and in this context the Fe-B system, which has been evaluated by various researchers [45–48], with some providing thermodynamic calculations [46,49–51]. The primary difference among models lies in the site of boron dissolution within steel phases. Boron can be considered the smallest substitutional element [15,49], the largest interstitial element [50], or a mixed substitutional-interstitial element in bcc-Fe and fcc-Fe [45,52–54]. Raghavan [55] gives a brief review of experimental data for the extension to the Fe-B-N system. Tomashik [56] provides an up-to-date review of the available data and claims that ternary phases do not occur. Fountain and Chipman [57] demonstrate a decrease in nitrogen solubility with increasing boron content. The Fe-B, Fe-B-N, and Fe-B-C system has previously been evaluated and in part re-optimized by the present author [58], determining the influence of boron on the solvus temperature of BN, and the liquidus and solidus temperatures of the system. The descriptions of these systems are taken as suggested by Führer et al. [58]. The description of the Al-N system is taken from Saunders [59]. In a recent publication [27], the Fe-Al-N-C system is reassessed for the Fe-rich corner in dilute systems, with particular emphasis on accurate thermodynamic description of the AlN phase. Ti-carbonitrides, which are classified as part of the MX-class [60], are modeled as a FCC solid solution. The model parameters are defined in such a way as to allow for the separation between the austenitic steel phase and the carbo-nitride (MX) phase in the Fe-base system. The multi-component model is expressed as (Fe, Ti)(C, N, Va) within the compound energy formalism.

The binary subsystem description for Ti-C is adopted from the Matcalc steel database [61]. The parameters for Fe-Ti and Ti-N are taken from Lee [16]. In the context of the simulation, the ternary subsystem that has been determined to be of major importance is composed of Fe-Al-C and Fe-Al-N, as delineated by Chin et al. and Hillert and Jonsson [62]. The ternary extension to Fe-Ti-C system is taken from Jacob et al. [60]. Table 1 provides a summary of the subsystems used for describing the Fe-C-N-Al-B-Ti system.

**Table 1.** Summary of binary and ternary subsystems used in the present study.

| System | Citation |
|---|---|
| Fe-C | Gustafson [13] |
| Fe-N | Frisk [14] |
| Fe-B, Fe-B-C, Fe-B-N | Miettinen and Vassilev [63], Führer et al. [58] |
| Fe-Al | Seierstein [64], Chin et al. [17] |
| Fe-Ti | Lee [16] |
| Ti-N | Lee [16] |
| Ti-C | Povoden-Karadeniz [61] |
| Al-Ti | Witusiewicz et al. [65] |
| Al-N | Saunders [59], Führer et al. [27] |
| Al-C | Chin et al. [17] |
| Fe-Al-C | Chin et al. [17] |
| Fe-Al-N | Hillert and Jonsson [62] |
| Fe-Ti-C | Povoden-Karadeniz [61] |

### 2.3. *Gibbs Energy Description of the BN-Phase*

The solubility product, K, defines the concentration product between B and N in weight percent, where BN is dissolved in the steel matrix at a given temperature. Solubility products are then plotted logarithmically, log(*K*), as a function of temperature. The lower the log(*K*), the lower the respective concentrations of B and N at which BN is completely dissolved.

Due to decreasing element concentrations with decreasing log(*K*) values, a higher sensitivity of the experimental detection method is required. In general, conducting an experimental investigation of the slope of the solubility product at low concentrations poses a significant challenge. Conversely, BN precipitation occurs at higher concentrations of B and N than given by K. It is self-evident that log(*K*) is typically lower at lower temperatures: the phase boundary between the steel matrix phase and the precipitate at low temperatures is usually at more dilute compositions than at higher temperatures. As illustrated in Figure 1, a comparison is presented between the experimentally evaluated solubility products [57,66,67] (full lines) and the calculated ones [68–70] (dotted lines) in the austenite.

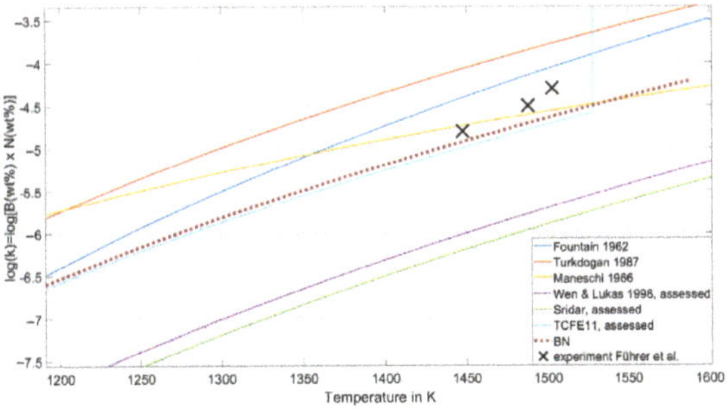


**Figure 1.** Full and dashed lines show a comparison of solubility products log(*K*) = log[B(wt.-%)·N(wt.-%)] in the austenite from experiments from the literature, adapted from Refs. [57,66,67], and assessed thermodynamic descriptions, adapted from Refs. [68–70], respectively. The red dotted line shows the result of the newly assessed description. Black crosses are the experimental results from Führer et al. adapted from Ref. [58]. (For the interpretation of the references to color in this figure legend, the reader is referred to the Web version of this article).

The available descriptions can be divided into two groups, with higher and lower solubility product values. Führer et al. [58] provide a comprehensive overview of the substantial discrepancy between the two groups. Importantly, the groups are defined by their approach to either correctly describing thermodynamic quantities, such as enthalpy of formation $\Delta H_f$, entropy $S_{298}$, and heat capacity $C_{p298}$, or correctly depicting solubility limits. Usually, microscopic dissolution observation should be consistent with the thermodynamic properties of a phase, controlling this behavior, as it could be seen for AlN [27]. However, the authors [58] have demonstrated that, in order to accurately depict the phase stability and thermodynamic quantities of the BN phase, for consistency between dissolution and thermodynamic properties, it is imperative to take into account the pronounced segregation tendency of B atoms, in conjunction with the Calphad-type thermodynamic description. In this work, an alternative Gibbs energy description of the BN phase is presented. The objective of this description is to provide a comprehensive characterization of the experimentally accessible solubility products, which involves the pronounced segregation tendency of B atoms, illustrated in Equation (1),

$$G_{Al:N}^{AlN} = -222000 + 45{\cdot}T + 7{\cdot}T{\cdot}ln(T) - 0.002{\cdot}T^2 - 3e^{-7}{\cdot}T^3 + GHSER_N + GHSER_B \quad (1)$$

It should be noted that this parameter proposal inevitably leads to different predicted thermodynamics of the BN phase. This results in deviations from the experimental values of Dworkin et al. [71] for enthalpy of formation ($\Delta H_f$), entropy ($S_{298}$), and heat capacity ($C_{p298}$) of approximately 25 $kJmol^{-1}$, 4 $Jmol^{-1}K^{-1}$, and < 0.1 $Jmol^{-1}K^{-1}$, respectively. The BN phase is regarded as fully stoichiometric (B)(N). For a more thorough examination of the BN phase description and a deeper discussion about the pertinent subsystems that contain boron, please refer to the extant literature [58,63].

## 3. Materials and Sample Characterization

### 3.1. Composition

In the present study, we used the Fe-Al-B-C-N alloying system with 4 and 5 variations in the B and Al contents from 0 to 0.083 wt.-% and from 0 to 0.4 wt.-%, respectively. In the Fe-Al-B-Ti-C-N system, the Ti microalloying content is varied from 0 to 0.042 wt.-% in four variations. The alloying of the material was carried out with high precision in a vacuum induction melting furnace, followed by a casting and forging process and a homogenizing annealing treatment at 1000 °C for ten hours in a vacuum furnace. The resulting composition of the elements C, Al, B, Ti, Mn, and S was measured by optical emission spectroscopy (OES) Thermofisher Type 4460 (Thermo Fisher Scientific, Waltham, MA, USA) and the N fraction by the LECO method [72,73] (Leco ON 736, LECO Corporation, St. Joseph, MO, USA) (see Table 2) with an uncertainty of five ppm and seven ppm, respectively. The numbers in the sample designations indicate the boron fraction in wt.-ppm and the Al and Ti fractions in tens of wt.-ppm. The sample geometries were cylindrical with Ø 5 mm and a height of 10 mm for EDS particle analysis. The stability of precipitates, e.g., AlN, in which Al can form oxides, can be influenced to a significant degree by the presence of impurities. Thus, particularly in the case of phase evaluation in microalloying steels, the utilization of vacuum induction melting for sample preparation is mandatory. It has been demonstrated to result in a significant reduction in impurity levels (see Table 2), with oxygen levels measured in parts per million (wt.). Due to the extremely low impurity levels, the influence of impurity elements on the stability of AlN, BN, and TiN is considered negligible.

**Table 2.** Composition of the alloys produced using vacuum induction melting (wt..-%). Numbers in the sample labeling present boron in wt.-ppm and Al and Ti in tens of wt.-ppm.

| | **Mass Fraction in %, Fe = bal.** | | | | | | | |
|---|---|---|---|---|---|---|---|---|
| Alloy | C | N | Al | B | Ti | Mn | S | O |
| | | | Fe-C-N-Al-B-B-variation | | | | | |
| Al29B0 | 0.51 | 0.0076 | 0.029 | 0.0002 | 0.000 | 0.123 | 0.006 | 0.001 |
| Al29B22 | 0.50 | 0.0073 | 0.029 | 0.0022 | 0.000 | 0.123 | 0.006 | 0.001 |
| Al36B44 | 0.50 | 0.0070 | 0.036 | 0.0044 | 0.000 | 0.123 | 0.007 | 0.001 |
| | | | Fe-C-N-Al-B-Al-variation | | | | | |
| B21Al0 | 0.50 | 0.0070 | 0.009 | 0.0021 | 0.000 | 0.115 | 0.0070 | 0.001 |
| B19Al35 | 0.49 | 0.0079 | 0.035 | 0.0019 | 0.000 | 0.114 | 0.008 | 0.001 |
| B21Al87 | 0.49 | 0.0083 | 0.087 | 0.0021 | 0.000 | 0.114 | 0.008 | 0.001 |
| B21Al180 | 0.50 | 0.0086 | 0.180 | 0.0021 | 0.000 | 0.117 | 0.008 | 0.001 |
| B21Al416 | 0.49 | 0.0086 | 0.416 | 0.0021 | 0.000 | 0.117 | 0.009 | 0.001 |
| | | | Fe-C-N-Al-B-Ti-Ti-variation | | | | | |
| Al36B25Ti0 | 0.51 | 0.0069 | 0.036 | 0.0025 | 0.000 | 0.131 | 0.007 | 0.001 |
| Al38B36Ti20 | 0.50 | 0.0052 | 0.038 | 0.0036 | 0.020 | 0.131 | 0.007 | 0.001 |
| Al42B38Ti30 | 0.50 | 0.0045 | 0.042 | 0.0038 | 0.030 | 0.131 | 0.007 | 0.001 |
| Al44B39Ti42 | 0.50 | 0.0042 | 0.044 | 0.0039 | 0.042 | 0.130 | 0.007 | 0.001 |

## 3.2. Methods

### 3.2.1. Microstructure Analysis

The analysis of the resulting microstructures, focusing on AlN, BN, and TiN precipitates, was conducted using a ZEISS SIGMA scanning electron microscope (SEM) (carl zeiss microscopy gmbh, Oberkochen, Germany) equipped with a field emission gun operating at an electron beam voltage of 10 kV and a working distance of 10 mm. The chemical analysis was facilitated by EDS, while wavelength-dispersive spectroscopy (WDS) was employed for elements B and N with a low ordering number. The EDS detector utilized was an Octane elect super with a 70 mm$^2$ detector size, while the WDS spectrometer was an EDAX Lambdaplus WDS [74] model (EDAX Inc., Mahwah, NJ, USA). The backscatter electron mode (BSE) in the SEM offers a favorable mass contrast for the precipitates, given their comparatively low atomic masses relative to Fe. For good statistical power, either 50 sections were observed manually in the SEM or computer-aided particle analysis was used. The former austenite grain size was revealed by using the etching reagent PM1998 [75]. The $T_{sol}$ of the precipitates of the alloys from Table 2 were evaluated by annealing the samples at different temperatures in a Bähr dilatometer type 805 under a vacuum atmosphere with alteration steps of 30 °C for 30 min and then quenching them down to room temperature to "freeze" the microstructure. The temperature curve of the heat treatments is shown in Figure 2. A temperature resolution of 30 K was selected as a compromise between high resolution and the minimization of necessary experiments. It should be noted that unless otherwise specified, all heat treatments are assumed to have undergone heating and cooling rates of 70 K/s and 100 K/s, respectively.

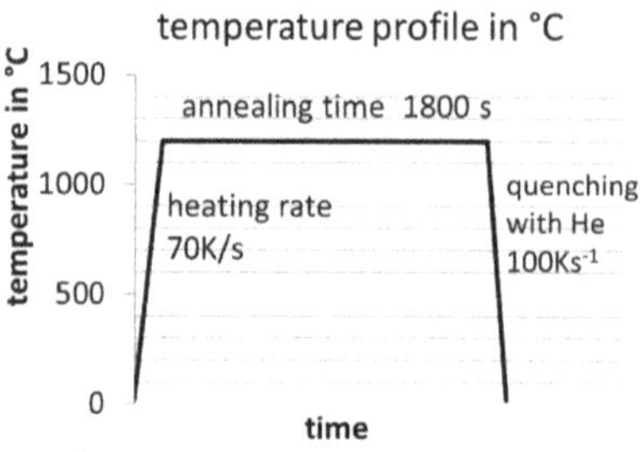


**Figure 2.** Applied temperature–time profile in a Bähr dilatometer type 805 under vacuum conditions for determining $T_{sol}$ of the nitrides.

3.2.2. EDS—Particle Analysis

In addition to the manual analysis of precipitates in the SEM, the computer-aided EDS particle analysis method (PA) [27] was employed. The analysis was conducted using a JEOL JSM-7800F scanning electron microscope (JEOL, Akishima, Japan) with a field emission gun at an operation voltage of 12 kV and an "Oxford X-MAXN" EDS detector (Oxford Instruments, Abingdon, UK) with a 50 mm$^2$ detector size for the computer-aided EDS particle analysis (PA). The primary advantage of PA was the acquisition of substantial statistical data concerning precipitates, including their distribution at interfaces, mean particle size and distribution, and phase fractions. Notably, PA effectively addresses the challenge posed by small and sparse precipitates, particularly those of AlN [27]. The evaluation of particles was conducted in accordance with the methodology outlined by Führer et al. [27]. The salient characteristics of this approach are as follows: the lower detection limit is 60 nm, and the determination of particles is facilitated by their chemical composition. However, this approach is encumbered by the difficulty of detecting light elements with EDS, which introduces a degree of uncertainty. In contrast, the substantial expanse of the investigative domain, in conjunction with the implementation of automated evaluation methodologies, facilitates the consideration of a considerable array of precipitates. Consequently, this approach engenders the generation of reliable statistics concerning precipitate distributions. The particles are then identified using certain composition boundaries, as outlined in Table 3. For AlN and MnS, the recommendation for the selection limit proposed by Führer et al. [27] is adhered to. The presence of BN is indicated by the detection of boron or nitrogen. The solubility of other elements is minimal, and thus, the detection of BN is a reliable indicator of the presence of these elements. BN contains light elements but possesses a substantial diameter, a property that enables its detection through particle analysis. Given the classification of TiN as a carbonitride, the high solubility of carbon must be considered. Given the fact that carbon and nitrogen are light elements and titanium is more readily detected with EDS, the detection of titanium is sufficient for marking a particle as TiN. Upon the satisfaction of all criteria, the spatial coordinates, mean ECD, surface fraction, and precipitate size distribution of the particles are determined.

**Table 3.** Selection limits for precipitates AlN, BN, TiN, MnS, and $Al_2O_3$ after particle analysis. Boundary conditions for elements are given in wt.-%.

| Type | | | | | | | | | |
|---|---|---|---|---|---|---|---|---|---|
| | ECD [nm] | Al [wt.-%] | B [wt.-%] | N [wt.-%] | Ti [wt.-%] | C [wt.-%] | Mn [wt.-%] | S [wt.-%] | O [wt.-%] |
| AlN | >60 | >2 | <0.1 | >0 | <0.1 | <3 | <3 | <0.1 | <0.1 |
| BN | >60 | <0.1 | >2 | >0 | <0.1 | <3 | <3 | <0.1 | <0.1 |

**Table 3.** *Cont.*

| Type | | | | | | | | | |
|---|---|---|---|---|---|---|---|---|---|
| | ECD [nm] | Al [wt.-%] | B [wt.-%] | N [wt.-%] | Ti [wt.-%] | C [wt.-%] | Mn [wt.-%] | S [wt.-%] | O [wt.-%] |
| TiN | >60 | <0.1 | <0.1 | >0 | >2 | >0 | <3 | <0.1 | <0.1 |
| MnS | >60 | <3 | <3 | <3 | <3 | <3 | >3 | >0.5 | <0.1 |
| $Al_2O_3$ | >60 | >2 | <3 | <3 | <3 | <3 | <3 | <0.1 | >1 |

# 4. Results

## *4.1. Initial Microstructure*

Figure 3 shows the microstructure of the alloy with different B fractions after annealing in a vacuum furnace at 1000 °C for 10 h, followed by cooling to room temperature at 90 K/min and etching with Nital. The results clearly demonstrate a correlation between an increase in grain size and an increase in B fraction between 22 and 44 wt.-ppm boron alloying. This phenomenon has been well-documented in the literature concerning B-alloyed steel and is attributed to the absence of AlN, which is known for its strong grain boundary pinning force, due to the formation of the more stable BN. Conversely, BN is too coarse to have any influence on grain growth [7]. Figure 3 reveals no discernible influence on the grain size between parts a and b for 0 and 22 wt.-ppm. As demonstrated in Table 4 and Figure 4 experimental and thermodynamic equilibrium simulation results, respectively, it is shown that AlN is stable at 1000 °C for alloys Al29B0 and Al29B22 and thus effectively retards grain growth. This indicates that the correct prediction of phase relations is highly relevant for interpretations of the according microstructure.

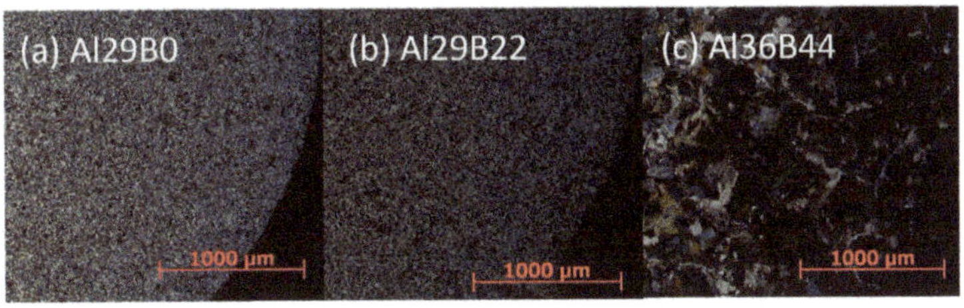


**Figure 3.** Microstructure of the samples after homogenization annealing in the vacuum furnace at 1000 °C for 10 h and subsequent cooling to room temperature with 90 K/min for the alloys with varying B fraction. Etching agent: Nital 5%. Magnification: ×25.

**Table 4.** Experimental and computed $T_{sol}$ of AlN and BN in the Fe-C-N-Al-B system.

| Alloy | Al | B | Ti | N | $T_{sol}$ AlN | $T_{sol}$ BN | $T_{sol}$ AlN | $T_{sol}$ BN |
|---|---|---|---|---|---|---|---|---|
| | in wt.-% | | | | experimental in °C | | simulative in °C | |
| Al29B0 | 0.029 | 0.0002 | 0.000 | 0.0076 | 1150 ± 30 | - | 1187 | - |
| Al29B22 | 0.029 | 0.0022 | 0.000 | 0.0073 | 1150 ± 30 | 1200 ± 15 | 1165 | 1235 |
| Al36B44 | 0.036 | 0.0044 | 0.000 | 0.0070 | 970 ± 40 | 1215 ± 15 | 1122 | 1280 |
| B21Al0 | 0.009 | 0.0021 | 0.000 | 0.0070 | - | 1230 ± 15 | - | 1230 |
| B19Al35 | 0.035 | 0.0019 | 0.000 | 0.0079 | 1180 ± 20 | 1215 ± 20 | 1195 | 1220 |
| B21Al88 | 0.088 | 0.0021 | 0.000 | 0.0083 | 1250 ± 20 | 1110 ± 20 | 1280 | - |
| B21Al180 | 0.180 | 0.0021 | 0.000 | 0.0086 | 1350 ± 50 | - | 1370 | - |
| B21Al416 | 0.416 | 0.0021 | 0.000 | 0.0086 | (1450 ± 50) | - | - | 1460 |

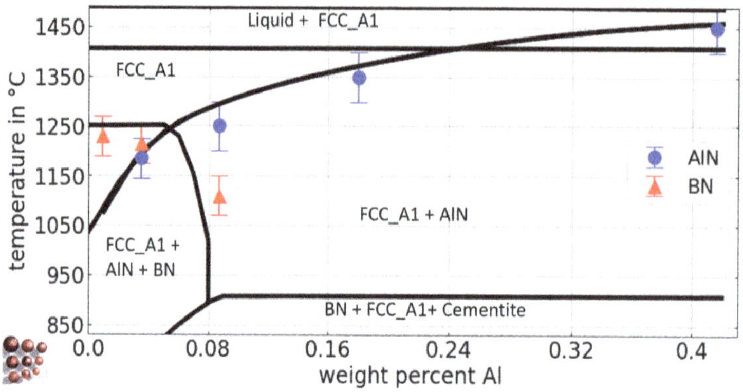


**Figure 4.** Calculated pseudo-binary Fe-Al phase diagram with 0.5 wt.-% C, 79 wt.-ppm N, and 21 wt.-ppm B for the austenite region of the Fe-C-N-Al-B system, ME-Fe_CDLIPE2.0.2.tdb. The blue circles and the red triangles mark the experimental results with error bars for the $T_{sol}$ of AlN and BN for the alloys B21Al0, B19Al35, B21Al88, B21Al180, and B21Al416, respectively. (For the interpretation of the references to color in this figure legend, the reader is referred to the Web version of this article).

### *4.2. Interrelation of AlN and BN Within Fe-C-N-Al-B*

#### 4.2.1. Solvus—Temperature of AlN and BN

In Table 4, the experimental results of $T_{sol}$ for AlN and BN precipitates using EDS/WDS analysis after annealing at different temperatures are presented, considering various Al and B concentrations, along with the associated range of error. The assessment of $T_{sol}$ for AlN with low Al fractions and varying B content, i.e., alloys Al29B0, Al29B22, and Al36B44, poses significant challenges due to the low phase fraction and small diameter of AlN [10,27]. Utilizing the outcomes derived from EDS particle analysis and acknowledging the microstructure that has undergone homogenization annealing at 1000 °C (Figure 3b,c), it can be deduced that $T_{sol}$ of alloy Al36B44 is approximately 970 °C. This determination is substantiated by the considerable augmentation in the affected grain size that has been observed following homogenization annealing, as evidenced by the comparative analysis in Figure 3b,c. The solubility of BN in the alloys with varying B fraction ranges from 1200 °C to 1215 °C, as BN is present at 1200 °C but not at 1235 °C or higher temperatures, in any of the investigated alloys. In alloys with a fixed B content and varying Al fractions, a decrease in the $T_{sol}$ of BN is observed, with complete suppression of BN formation at concentrations greater than 0.9 wt.-%. $T_{sol}$ of AlN at temperatures above 1300 °C (alloys B21Al180, B21Al416) is associated with a higher degree of uncertainty, primarily due to the challenges associated with conducting heat treatment experiments at such elevated temperatures, particularly the difficulty of achieving stable connection of the thermocouple on the sample surface.

For alloys B21Al180 and B21Al416, we therefore employ thermodynamic equilibrium simulation in order to discuss the accuracy of experimentally determined $T_{sol}$. Figure 5 shows an SE image (a) and the EDS mapping of Al (b) and Mn (c) for alloy B21Al88 after annealing at 1100 °C for 1800 s. The concentrations of Al in part b reveal the presence of AlN particles with a size in the range of several hundred nanometers.

Figure 6 shows the chemical analysis by EDS of AlN and MnS precipitates located along each other in the alloy B21Al88 after annealing at 1180 °C for 1800 s. The EDS mappings (a, b, c, d) in Figure 6 give more details on the distribution of Al, N, Mn and S atoms, showing the actual size of the precipitate and the positioning of the two precipitates along each other. A close nesting between MnS and AlN is evident. AlN precipitates are found in the size range of approx. 100 nm, which is close to the lower resolution limit for PA up to 1 μm. A clear dependence of the AlN particle size on the Al fraction in the alloy is observed, with AlN particles being smaller than 300 nm in alloy B21Al35 and reaching particle diameters larger than 1 μm in alloy B21Al88. Co-precipitation of AlN and MnS in microalloyed steel has been reported by [27,76] and is also observed in this study, see Figure 6. AlN precipitates in the size of 100 nm are not found with a solubility of Mn, as shown in Figure 5.

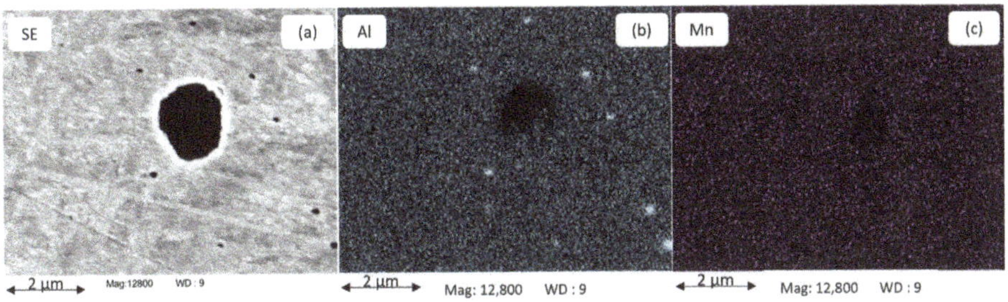


**Figure 5.** SE image (**a**) and chemical analysis mapping for Al and Mn (**b**,**c**) utilizes EDS to examine AlN precipitates in the size range of several hundred nanometers in alloy B21Al88 after annealing at 1100 °C for 1800 s.

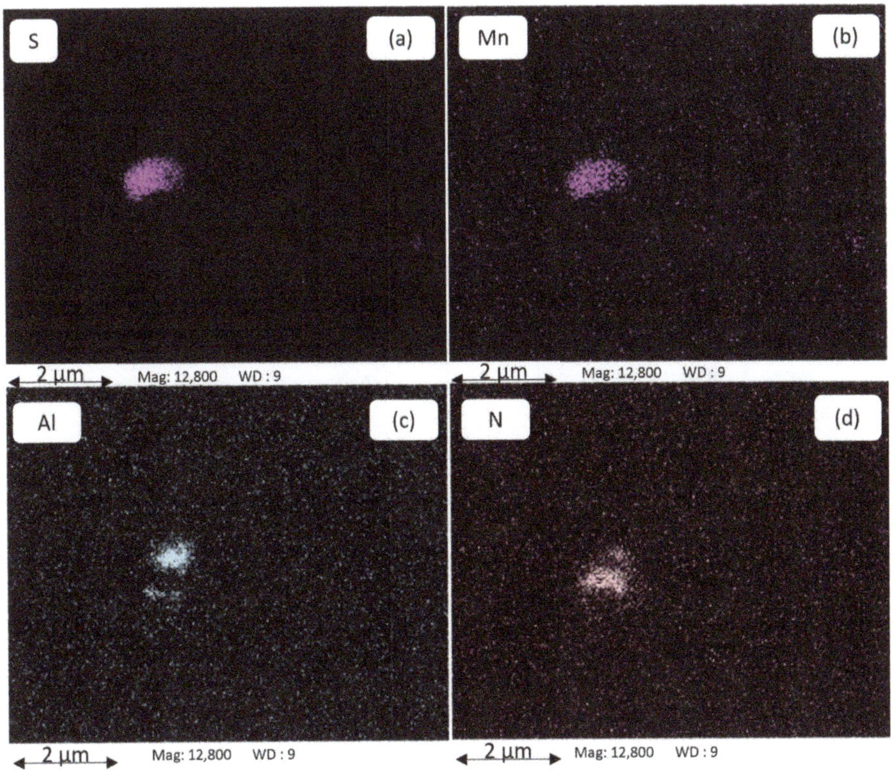


**Figure 6.** Chemical analysis mapping for S, Mn, Al, and N (**a**–**d**) utilizes EDS to examine an AlN precipitate alongside a MnS precipitate in alloy B21Al88 after annealing at 1180 °C for 1800 s.

The occurrence of two distinct peaks in the distribution of AlN diameters is evident. As demonstrated in Figure 5, the formation of small AlN particles with diameters of approximately 100 nm occurs during processing at lower temperatures. In contrast, the formation of larger AlN particles occurs at higher temperatures, with the nucleation being initiated at these elevated temperatures. As the aluminum composition increases, the solvus temperature of the aluminum nitride (AlN) also increases. Consequently, the size of the particles increases as well, due to the higher diffusivities that occur at higher temperatures.

Figure 7 shows a corresponding EDS/WDS analysis of a BN precipitate (b), typically located at an austenite grain boundary, in alloy Al36B44 after annealing for 1800 s at 1200 °C. The spectrum (a) from the WDS analysis shows the presence of N.

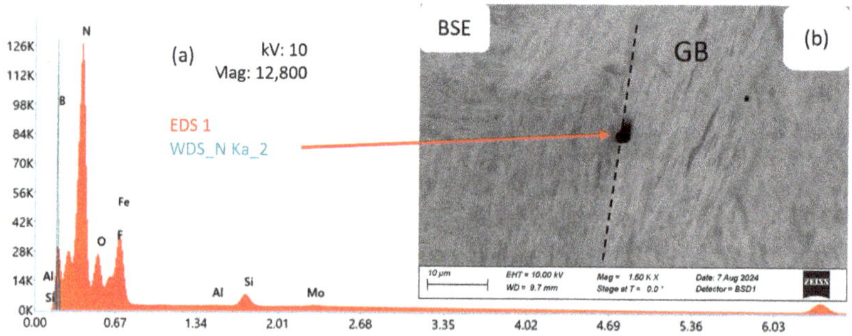


**Figure 7.** SEM-BSE image of a BN precipitate (**b**) at a grain boundary (GB). Chemical analysis of a BN precipitate using EDS and WDS (**a**) for alloy Al36B44 after annealing for 1800 s at 1200 °C. B is detected using WDS.

From the analysis in the SEM, it can be concluded that BN precipitates appear as roundish features in the range of 1–5 μm and are predominantly nucleated at grain boundaries, as reported in the literature [7,31].

The protection of boron from nitrogen by excess aluminum is controversial. According to the review by Sharma et al. [7], it is believed that AlN dissolves at high temperatures, allowing the N to recombine with B. However, this study shows that AlN remains stable at high temperatures. In a system with 21 wt.-ppm B and 83 wt.-ppm N, an Al fraction greater than 0.09 wt.-% is sufficient to completely suppress BN. As a rule of thumb for the preferential precipitation of AlN over the formation of BN, an Al/N ratio of 20 is found for a B fraction of 21 wt.-ppm.

#### 4.2.2. Pseudo-Binary Fe-Al and Fe-B Phase Diagram Within the Fe-N-C-Al-B System

As illustrated in Figure 4, the pseudo-binary Fe-Al diagram for the system Fe-C-N-Al-B with 0.5 wt.-% C, 79 wt.-ppm N, and 21 wt.-ppm B for the austenite region is calculated using the thermodynamic database ME-Fe_CDLIPE2.0.2.tdb [18].

The blue circles and red triangles represent the experimental results from Section 4.2.1 for the $T_{sol}$ of AlN and BN, respectively. The $T_{sol}$ of BN for the alloys B21Al0, B19Al35, and B21Al88 lies within a narrow range, as evidenced by the experimental results. From a thermodynamic perspective, the higher affinity of boron for nitrogen compared to aluminum [8,77] elucidates the amount of aluminum necessary to restrict the solubility of BN in austenite [7], with approximately 0.09 wt.-% required to entirely impede BN formation. Figure 8 shows the pseudo-binary Fe-B diagram for the Fe-rich side of the system Fe-C-N-Al-B with 0.5 wt.-% C, 79 wt.- ppm N, and 0.03 wt.-% Al for the austenite region. These calculations are derived from the thermodynamic database ME-Fe_CDLIPE2.0.2.tdb [18].

The blue circles and red triangles represent the experimental results from Section 4.2.1 for $T_{sol}$ of AlN and BN, respectively.

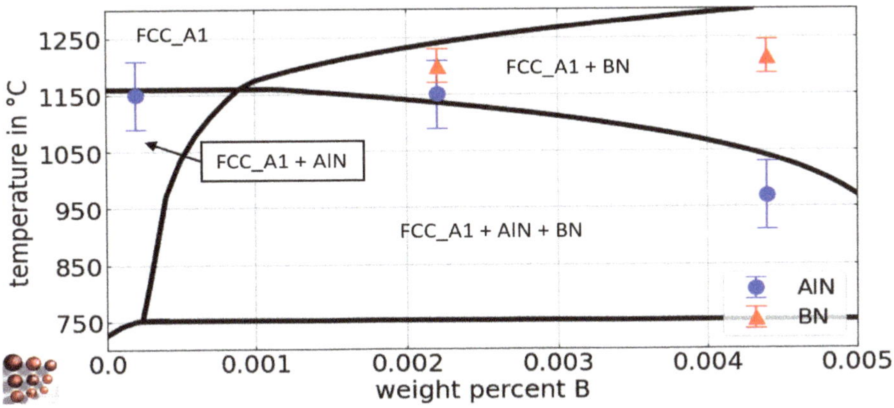


**Figure 8.** Calculated pseudo-binary Fe-B phase diagram with 0.5 wt.-% C, 79 wt.-ppm N, and 0.03 wt.-% Al for the austenite region of the Fe-C-N-Al-B system, ME-Fe_CDLIPE2.0.2.tdb. The blue circles and the red triangles mark the experimental results with error bars for the $T_{sol}$ of AlN and BN for the alloys Al29B0, Al29B22, and Al36B44. (For the interpretation of the references to color in this figure legend, the reader is referred to the Web version of this article).

As the boron fraction increases, the stability of BN begins to manifest at approximately 4 wt.-ppm B. The presence of competing phases, AlN and BN, leads to a reduction in AlN precipitates with an increase in boron content, owing to the formation of BN. At B concentrations greater than 50 wt.-ppm, AlN is completely dissolved.

This work definitively shows that AlN and BN are competing phases. An increase in either aluminum or boron content favors the formation of the corresponding phase, AlN or BN. The BN phase is generally more stable than AlN due to the strong affinity of boron for nitrogen, depending on the relative concentrations of Al, B, and N.

### 4.3. *Interrelation of AlN, BN, and TiN Within Fe-C-N-Al-B-Ti*

#### 4.3.1. Experiment

In Table 5, the experimental findings concerning the $T_{sol}$ of the precipitate phases AlN, BN, and TiN for the alloys exhibiting variable titanium concentrations are presented. The data shows the strong affinity of Ti for N, as evidenced by the observation that a mere 0.03 wt.-% Ti is enough to entirely impede the formation of AlN and BN. This phenomenon, consequently, facilitates the advantageous effect of safeguarding B for the desired "boron-effect" outcome. There is no evidence of mixing between the AlN, BN, and TiN phases in the EDS mapping. The determination of $T_{sol}$ for TiN is achieved through the integration of thermodynamic equilibrium simulation with experimental evaluation. In the case of alloy Al36B25Ti0, no TiN is detected, whereas in all other alloys containing Ti, TiN is found. The high solubility temperature of TiN, which is approximately 1430 °C or higher, renders dissolution experiments in the dilatometer unfeasible. In Figure 9, an exemplary corresponding EDS analysis of a TiN precipitate in alloy Al42B38Ti30 after annealing at 1300 °C for 1200 s is shown, yielding a spectrum (a) that substantiates the presence of Ti and N. The EDS mappings (b, c, d, and e) provide further insight into the distribution of the Ti, N, Fe, and C atoms, respectively, and demonstrate the actual size of the precipitate.

The size of the detected TiN precipitates ranges from 100 nm (lower detection limit) to several μm.

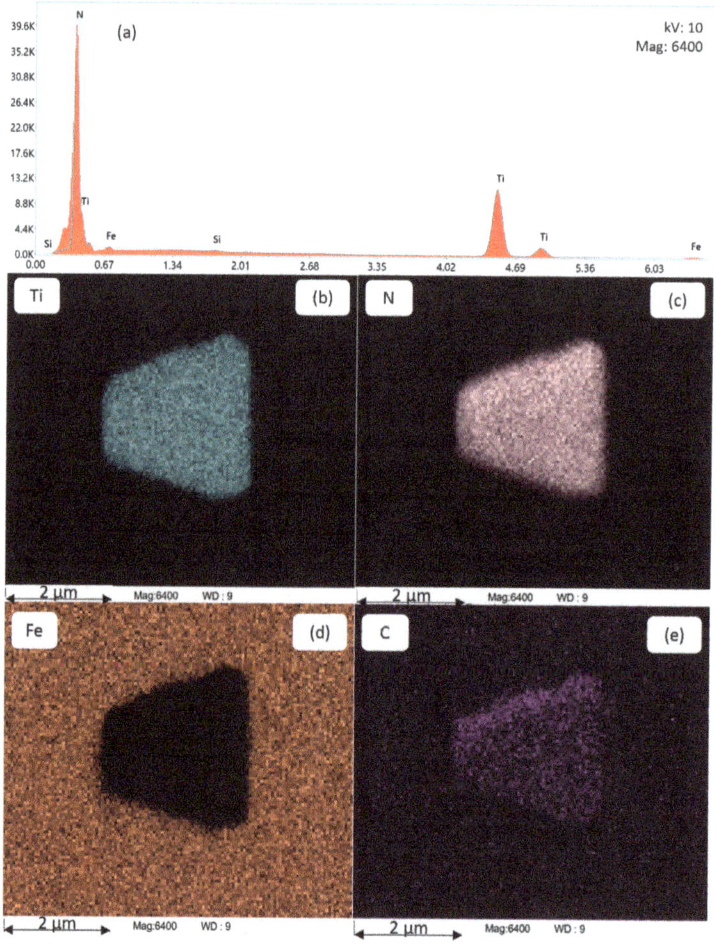


**Figure 9.** Chemical analysis of a TiN precipitate using EDS (**a**) in alloy Al42B38Ti30 after annealing at 1300 °C for 1200 s. The EDS mappings in (**b–e**) show the distribution of Ti, N, Fe, and C atoms, respectively.

**Table 5.** Experimental and computed $T_{sol}$ of AlN, BN, and TiN in the Fe-C-N-Al-B-Ti system. The values for TiN are taken from thermodynamic equilibrium simulation.

| Alloy | Al | B | Ti | N | $T_{sol}$ AlN | $T_{sol}$ BN | $T_{sol}$ AlN | $T_{sol}$ BN | $T_{sol}$ TiN |
|---|---|---|---|---|---|---|---|---|---|
| | | in wt.-% | | | experiment in °C | | simulation in °C | | |
| Al36B25Ti0 | 0.036 | 0.0025 | 0.000 | 0.0069 | 1140 ± 20 | 1230 ± 20 | 1170 | 1240 | - |
| Al38B36Ti20 | 0.038 | 0.0036 | 0.020 | 0.0052 | - | 1120 ± 20 | - | 1100 | 1430 ± 20 |
| Al42B38Ti30 | 0.042 | 0.0038 | 0.030 | 0.0045 | - | - | - | - | 1440 ± 20 |
| Al44B39Ti42 | 0.044 | 0.0039 | 0.042 | 0.0042 | - | - | - | - | 1450 ± 20 |

As illustrated in Figure 10, the results of the particle analysis for alloy Al42B38Ti30 after annealing at 1000 °C for 10,000 s indicate the presence of TiN and MnS. The blue and red dots denote the positions of these two phases, respectively. The precipitate size in the plot is multiplied by 20 to enhance the visibility of the particles. The absence of BN and AlN in the present measurement further suggests that AlN and BN are not stable within the alloy and annealing temperature under investigation. The diameter of titanium nitride (TiN) is distributed over a wide range, from a mere hundred nanometers to several micrometers. The formation of large TiN precipitates is hypothesized to occur at elevated temperatures or primarily from the liquid phase [78]. Small TiN particles are thought to nucleate as secondary precipitates from the solid state at lower temperatures during the subsequent production process.

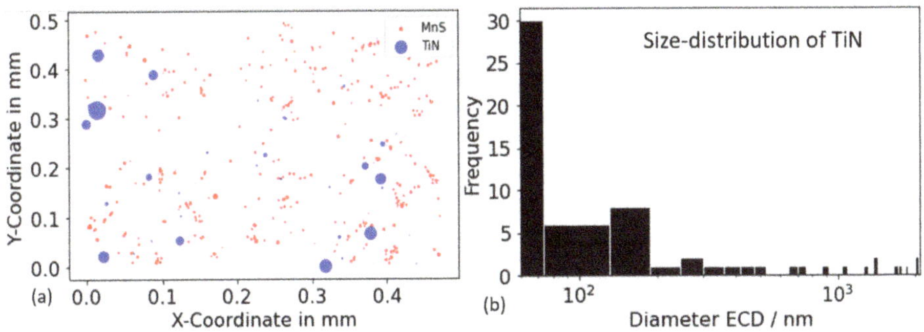


**Figure 10.** Results of particle analysis for alloy Al42B38Ti30 after annealing at 1000 °C for 10,000 s. (**a**) Position of the TiN (blue) and MnS (red) precipitates. The precipitate size in the plot is multiplied by 20 to make the particles more easily visible. (**b**) Distribution of the TiN precipitates diameter.

Figure 11 shows the resulting austenite grain size after annealing at 1000 °C for 500 s and 10,000 s, respectively. A profound bimodal grain size distribution can be observed, indicating the abnormal grain growth behavior. This behavior is explained by the absence of grain boundary pinning AlN precipitates. A similar effect is reported by Roy et al. [37] and Monschein et al. [35] in HSLA steel, where the dissolution of second-phase particles at higher soaking temperatures is reported to allow AGG. In the alloys without Ti alloying, AGG was not observed, further supporting the theory that the lack of AlN leads to AGG.

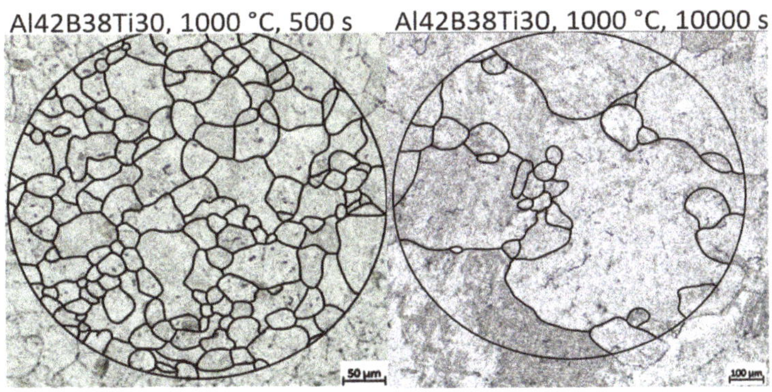


**Figure 11.** Comparison of the resulting austenite grain size for alloy Al42B38Ti30 after annealing at 1000 °C for 500 s and 10,000 s, respectively. Etching is performed with PM1998, adapted from Ref. [75]. Black lines are drawn to highlight the PAGB.

#### 4.3.2. Pseudo-Binary Fe-Ti Phase Diagram Within the Fe-N-C-Al-B-Ti System

Figure 12 shows a calculated pseudo-binary Fe-Ti diagram for the Fe-rich side with 0.5 wt.-% C, 79 wt.-ppm N, 0.03 wt.-% Al, and 36 wt.-ppm B for the austenite region of the Fe-C-N-Al-B-Ti system, using the thermodynamic database ME-Fe_CDLIPE2.0.2.tdb [18]. The blue circles, red triangles, and green stars represent the experimental results from Section 4.2.1 for the $T_{sol}$ of TiN, BN, and AlN, respectively. However, small discrepancies between simulation results and experimental data are observed, and they are attributed to experimental evaluation uncertainties, primarily due to resolution limits and the complex interactions with numerous degrees of freedom in the multicomponent system. In fact, the calibration of such interactions can be achieved, but we prefer to accept deviations due to the high doubt of the appropriate physical base for their modeling. The collective analysis of the simulation results indicates a reasonable agreement between the simulations and the experimental observations, suggesting a satisfactory overall outcome.

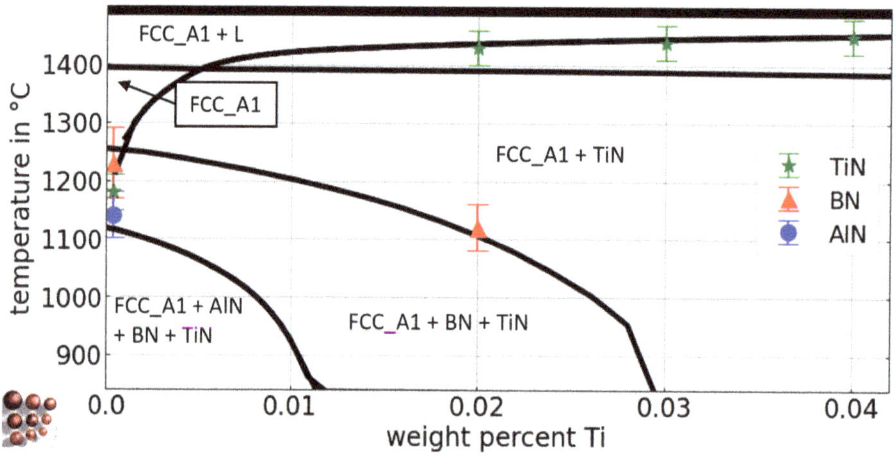


**Figure 12.** Calculated pseudo-binary Fe-Ti phase diagram with 0.5 wt.-% C, 79 wt.- ppm N, 0.03 wt.-% Al, and 36 wt.-ppm B for the austenite region of the Fe-C-N-Al-B-Ti system, ME-Fe_CDLIPE2.0.2.tdb. The blue circles, red triangles, and green stars mark the experimental results with error bars for the $T_{sol}$ of TiN, BN, and AlN for the alloys Al36B25Ti0, Al38B36Ti20, Al42B38Ti30, and Al44B39Ti42. (For the interpretation of the references to color in this figure legend, the reader is referred to the Web version of this article).

## 5. Summary

Our experimental investigation of high-purity vacuum induction-melted alloys confirms that Ti, B, and Al exhibit decreasing affinities to nitrogen, forming nitrides in microalloyed steel. In the competitive interaction with aluminum, in the Fe-C-N-Al-B system with approximately 80 wt-ppm N and 20 wt.-ppm B, increasing the Al content to 0.09 wt.-% effectively inhibits the formation of BN, in contrast to previous understanding of the negligible protective role of Al at high temperatures. AlN is found to heterogeneously nucleate on MnS. Titanium proves to be very effective in blocking the formation of other nitrides. The presence of 0.02 wt.-% Ti completely blocks the formation of AlN and shifts the $T_{sol}$ of BN by 150 °C. The addition of 0.03 wt.-% Ti is sufficient to suppress the formation of both BN and AlN, in steel with approximately 50 wt.-ppm N. By integrating available binary and ternary descriptions into a thermodynamic database, we predict the stabilities of the multicomponent system Fe-C-N-Al-B-Ti. A new Gibbs energy description of the BN phase is presented, which provides a comprehensive characterization of the experimentally accessible solubility products. Overall, considering the simulation results for nitride phase boundaries, the outcomes are satisfying, as reflected by a reasonable agreement between the simulations and experimental observations. This further confirms the control by the thermodynamics of binary and ternary subsystems, since the extension to quaternaries, quinaries, and the multi-component steel system studied has been considered ideal, and the applicability of the steel database, being optimized by new experimental data in the Fe-rich corner for dilute microalloyed steel is confirmed.

In alloys with even a small amount of Ti, abnormal grain growth behavior is observed. This inhomogeneous grain growth is attributed to the absence of AlN precipitates, which are known to retard grain boundary movement.

**Author Contributions:** Conceptualization, M.F. and E.P.-K.; methodology, M.F., C.S. and E.P.-K.; software, M.F. and E.P.-K.; investigation, M.F. and C.S.; writing—original draft preparation, M.F.; writing—review and editing, M.F., S.Z., C.S. and E.P.-K.; visualization, M.F.; supervision, E.P.-K. All authors have read and agreed to the published version of the manuscript.

**Funding:** This research was funded by Christian Doppler Forschungsgesellschaft in the framework of the CD-Laboratory of Interfaces and Precipitation Engineering (CDL-IPE).

**Data Availability Statement:** The original contributions presented in this study are included in the article. Further inquiries can be directed to the corresponding author.

**Acknowledgments:** The financial support from the Austrian Federal Ministry for Digital and Economic Affairs and the National Foundation for Research, Technology, and Development is gratefully acknowledged. The authors acknowledge TU Wien Bibliothek for financial support through its Open Access Funding Programme. The financial support, as well as the provision of sample material by voestalpine Forschungsservicegesellschaft, is gratefully acknowledged by the authors.

**Conflicts of Interest:** Author Sabine Zamberger was employed by the company Voestalpine Forschungsservicegesellschaft Donawitz GmbH. The remaining authors declare that the research was conducted in the absence of any commercial or financial relationships that could be construed as a potential conflict of interest.

## References

1. Baker, T.N. Microalloyed Steels. *Ironmak. Steelmak.* **2016**, *43*, 264–307. [CrossRef]
2. e Silva, A.C. Challenges and opportunities in thermodynamic and kinetic modeling microalloyed HSLA steels using computational thermodynamics. *Calphad* **2020**, *68*, 101720. [CrossRef]
3. Habu, R.; Miyata, M.; Tamukai, S.; Sekino, S. Improvement of Hardenability of Steel Containing Aluminum and Boron by Double Quenching. *Trans. Iron Steel Inst. Jpn.* **1983**, *23*, 176–183. [CrossRef]
4. Kawamura, K.; Otsubo, T.; Furukawa, T. Relationship between the Hardenability of Steel and the "Effective Boron" in Steel. *Trans. Iron Steel Inst. Jpn.* **1976**, *16*, 545–550. [CrossRef]
5. McMahon, C.J. The role of solute segregation in promoting the hardenability of steel. *Met. Trans. A* **1980**, *11*, 531–535. [CrossRef]
6. Wipp, D. Boron Influence on Microstructural Evolution and Mechanical Properties in Micro-Alloyed Carbon Steels: Precipitation and Segregation Behavior. Ph.D. Thesis, Technische Universität Wien, Vienna, Austria.
7. Sharma, M.; Ortlepp, I.; Bleck, W. Boron in Heat-Treatable Steels: A Review. *Steel Res. Int.* **2019**, *90*, 1900133. [CrossRef]
8. Wang, W.S.; Zhu, H.Y.; Sun, J.; Lei, J.L.; Duan, Y.Q.; Wang, Q. Thermodynamic analysis of BN, AlN AND TiN Precipitation in boron-bearing steel. *Metalurgija* **2019**, *58*, 199–202.
9. Baker, T.N. Titanium microalloyed steels. *Ironmak. Steelmak.* **2019**, *46*, 1–55. [CrossRef]
10. Wilson, F.G.; Gladman, T. Aluminium nitride in steel. *Int. Mater. Rev.* **1988**, *33*, 221–286. [CrossRef]
11. Gontijo, M.; Chakraborty, A.; Webster, R.F.; Ilie, S.; Six, J.; Primig, S.; Sommitsch, C. Thermomechanical and Microstructural Analysis of the Influence of B- and Ti-Content on the Hot Ductility Behavior of Microalloyed Steels. *Metals* **2022**, *12*, 1808. [CrossRef]

12. Kaufman, L.; Bernstein, H. *Computer Calculation of Phase Diagrams with Special Reference to Refractory Metals*; Academic Press Inc.: New York, NY, USA, 1970.
13. Gustafson, P. A thermodynamic evaluation of the Fe-C system. *Calphad* **1985**, 259–267. [CrossRef]
14. Frisk, K. A thermodynamic evaluation of the Cr-N, Fe-N, Mo-N and Cr-Mo-N systems. *Calphad* **1991**, *15*, 79–106. [CrossRef]
15. Miettinen, J.; Visuri, V.-V.; Fabritius, T.; Vassilev, G. Thermodynamic Description of Ternary Fe-B-X Systems. Part 7: Fe-B-C. *Arch. Met. Mater.* **2020**, *65*, 923–933. [CrossRef]
16. Lee, B.-J. Thermodynamic assessment of the Fe-Nb-Ti-C-N system. *Met. Mater. Trans. A* **2001**, *32*, 2423–2439. [CrossRef]
17. Chin, K.-G.; Lee, H.-J.; Kwak, J.-H.; Kang, J.-Y.; Lee, B.-J. Thermodynamic calculation on the stability of (Fe, Mn) 3AlC carbide in high aluminum steels. *J. Alloy. Compd.* **2010**, *505*, 217–223. [CrossRef]
18. Povoden-Karadeniz, E. (Ed.) Thermodynamic Database "ME-Fe_CDLIPE2.0.2.tdb. Unpublished work, 2025.
19. Rios, P.R.; da Fonseca, G.S. Grain Boundary Pinning by Particles. *Mater. Sci. Forum* **2010**, *638–642*, 3907–3912. [CrossRef]
20. Radis, R.; Kozeschnik, E. Kinetics of AlN Precipitation in Microalloyed Steel. *Model. Simul. Mater. Sci. Eng.* **2010**, *18*, 55003. [CrossRef]
21. Wriedt, H.A. The Al−N (Aluminum-Nitrogen) System. *Bull. Alloy. Phase Diagr.* **1986**, *7*, 329–333. [CrossRef]
22. Bruls, R.; Hintzen, H.; de With, G.; Metselaar, R.; van Miltenburg, J. The temperature dependence of the Grüneisen parameters of $MgSiN_2$, AlN and β-$Si_3N_4$. *J. Phys. Chem. Solids* **2001**, *62*, 783–792. [CrossRef]
23. Lu, X.-G.; Selleby, M.; Sundman, B. Assessments of molar volume and thermal expansion for selected bcc, fcc and hcp metallic elements. *Calphad* **2005**, *29*, 68–89. [CrossRef]
24. Sennour, M.; Esnouf, C. Contribution of Advanced Microscopy Techniques to Nano-Precipitates Characterization: Case of AlN Precipitation in Low-Carbon Steel. *Acta Mater.* **2003**, *51*, 943–957. [CrossRef]
25. Gao, N.; Baker, T.N. Austenite Grain Growth Behaviour of Microalloyed Al-V-N and Al-V-Ti-N Steels. *ISIJ Int.* **1998**, *38*, 744–751. [CrossRef]
26. Chen, Y.-L.; Wang, Y.; Zhao, A. Precipitation of AlN and MnS in Low Carbon Aluminium-Killed Steel. *J. Iron Steel Res. Int.* **2012**, *19*, 51–56. [CrossRef]
27. Führer, M.; Zamberger, S.; Povoden-Karadeniz, E. Experimental determination of AlN in microalloyed steel and thermodynamic analysis. *Calphad* **2024**, *88*, 102790. [CrossRef]
28. Massardier, V.; Guétaz, V.; Merlin, J.; Soler, M. Kinetic and Microstructural Study of Aluminium Nitride Precipitation in a Low Carbon Aluminium-Killed Steel. *Mater. Sci. Eng. A* **2003**, *355*, 299–310. [CrossRef]
29. Pearson, W.B.; Villars, P.; Calvert, L.D. *Pearson's Handbook of Crystallographic Data for Intermetallic Phases*; ASM International: Materials Park, OH, USA, 1986.
30. Suzuki, S.; Tanino, M. Role of Grain Boundary Segregation in Austenite Decomposition of Low-Alloyed Steel. *Mater. Sci. Forum* **2007**, *558–559*, 965–970. [CrossRef]
31. Sakuraya, K.; Okada, H.; Abe, F. Influence of Heat Treatment on Formation Behavior of Boron Nitride Inclusions in P122 Heat Resistant Steel. *ISIJ Int.* **2006**, *46*, 1712–1719. [CrossRef]
32. Jonsson, S. Assessment of the Fe-Ti-C system, calculation of the Fe-TiN system, and prediction of the solubility limit of Ti(C,N) in liquid Fe. *Met. Mater. Trans. B* **1998**, *29*, 371–384. [CrossRef]
33. ASM Handbook Committee. High-Strength Structural and High-Strength Low-Alloy Steels. In *Properties and Selection: Irons, Steels and High Performance Alloys*, 10th ed.; ASM Metals Handbook; ASM: Metals Park, OH, USA, 1990; Volume 1, pp. 403–423.
34. Najafkhani, F.; Kheiri, S.; Pourbahari, B.; Mirzadeh, H. Recent Advances in the Kinetics of Normal/Abnormal Grain Growth: A Review. *Arch. Civ. Mech. Eng.* **2021**, *21*, 29. [CrossRef]
35. Monschein, S.; Ragger, K.S.; Zügner, D.; Fasching, J.; Schnitzer, R. Influence of the Ti Content on the Grain Stability and the Recrystallization Behavior of Nb-Alloyed High-Strength Low-Alloyed Steels. *Steel Res. Int.* **2022**, *93*, 2200094. [CrossRef]
36. Su, C.; Zhao, G.; Xiao, H.; Lan, Y.; Huang, F. Abnormal Grain Growth of Hi-B Steel in the Secondary Recrystallization. *Met. Microstruct. Anal.* **2018**, *7*, 608–617. [CrossRef]
37. Roy, S.; Karmakar, A.; Mukherjee, S.; Kundu, S.; Srivastava, D.; Chakrabarti, D. Effect of starting microstructure on austenite grain sizes developed after reheating of HSLA steel. *Mater. Sci. Technol.* **2014**, *30*, 1142–1153. [CrossRef]
38. Lang, D.; Povoden-Karadeniz, E.; Schatte, J.; Knabl, W.; Clemens, H.; Primig, S. Thermodynamic evaluation of the Mo-rich corner of the Mo-Hf-C system including O impurities. *J. Alloy. Compd.* **2017**, *695*, 372–381. [CrossRef]
39. Saunders, N.; Miodownik, A.P. *CALPHAD (Calculation of Phase Diagrams): A Comprehensive Guide, Transferred to Digital Print*; Pergamon Materials Series: Pergamon, Turkey, 1998.
40. Dinsdale, A. SGTE Data for Pure Elements. *Calphad* **1991**, *15*, 317–425. [CrossRef]
41. Chipman, J. Thermodynamics and phase diagram of the Fe-C system. *Met. Trans.* **1972**, *3*, 55–64. [CrossRef]

42. Göhring, H.; Leineweber, A.; Mittemeijer, E. A thermodynamic model for non-stoichiometric cementite; the Fe–C phase diagram. *Calphad* **2016**, *52*, 38–46. [CrossRef]
43. Naraghi, R.; Selleby, M.; Ågren, J. Thermodynamics of stable and metastable structures in Fe–C system. *Calphad* **2014**, *46*, 148–158. [CrossRef]
44. Hallstedt, B.; Djurovic, D.; von Appen, J.; Dronskowski, R.; Dick, A.; Körmann, F.; Hickel, T.; Neugebauer, J. Thermodynamic properties of cementite. *Calphad* **2010**, *34*, 129–133. [CrossRef]
45. Van Rompaey, T.; Kumar, K.H.; Wollants, P. Thermodynamic optimization of the B–Fe system. *J. Alloy. Compd.* **2002**, *334*, 173–181. [CrossRef]
46. Chart, T.G. *CECA, Research Project No.7210, CA/3/303*; Comission des Communautès Europèennes: Brussels, Belgium, 1981.
47. Massalski, T.B. (Ed.) *Binary Alloys Phase Diagrams*, 2nd ed.; ASM International: Metals Park, OH, USA, 1990.
48. Kubaschewski, O. *Iron–Binary Phase Diagrams*; Springer: Berlin, Germany, 1982.
49. Hallemans, B.; Wollants, P.; Roos, J.R. Thermodynamic Reassessment and Calculation of the Fe-B Phase Diagram. *Int. J. Mater. Res.* **1994**, *85*, 676–682. [CrossRef]
50. Ohtani, H.; Hasebe, M.; Ishida, K.; Nishizawa, T. Calculation of Fe-C-B ternary phase diagram. *Trans. Iron Steel Inst. Jpn.* **1988**, *28*, 1043–1050. [CrossRef]
51. Kaufman, L.; Uhrenius, B.; Birnie, D.; Taylor, K. Coupled pair potential, thermochemical and phase diagram data for transition metal binary systems-VII. *Calphad* **1984**, *8*, 25–66. [CrossRef]
52. Lucci, A.; Della Gatta, G.; Venturello, G. On the Solubility of Boron in High-Purity Alpha-Iron. *Met. Sci. J.* **1969**, *3*, 14–17. [CrossRef]
53. Brodowsky, H.; Wernicke, H.-J. Activity coefficients and the $\alpha$-$\gamma$ transition lines in Fe-B alloys. *Calphad* **1984**, *8*, 159–162. [CrossRef]
54. Brown, A.; Garnish, J.D.; Honeycombe, R.W.K. The Distribution of Boron in Pure Iron. *Met. Sci.* **1974**, *8*, 317–324. [CrossRef]
55. Raghavan, V. B-Fe-N (boron-iron-nitrogen). *J. Phase Equilibria Diffus.* **1993**, *14*, 619–620. [CrossRef]
56. Tomashik, V. Boron—Iron—Nitrogen: Iron systems: Phase diagrams, crystallographic and thermodynamic data. *Landolt-Börnstein—Group IV Phys. Chem.* **2008**, 457–471. [CrossRef]
57. Fountain, R.W.; Chipman, J. *Solubility and Precipitation of Boron Nitride in Iron-Boron Alloys*; Union Carbide Metals Co.: Niagara Falls, NY, USA, 1962; Volume 224.
58. Führer, M.; Zamberger, S.; Shan, Y.; Helml, L.; Srikakulapu, K.; Schnitzer, R.; Kozeschnik, E.; Povoden-Karadeniz, E. *Influence of Boron Grain Boundary Segregation on the Thermodynamics of the Fe-C-B-N System*; unpublished manuscript; Institue of materials science and technology: TU Wien, Vienna, Austria, 2025.
59. Saunders, N. COST 507: Thermochemical database for light metal alloys. *Light Met.* **1998**, *2*, 23–27.
60. Jacob, A.; Sobotka, E.; Povoden-Karadeniz, E. Thermodynamic modeling of multicomponent MX phases (M = Nb, Ti, V; X = C, N) in steel. *Calphad* **2025**, *88*, 102795. [CrossRef]
61. Povoden-Karadeniz, E. mc_fe_v2.061. 2023. Available online: https://www.matcalc.at/images/stories/Download/Database/mc_fe_v2061.tdb (accessed on 23 August 2023).
62. Hillert, M.; Jonsson, S. An Assessment of the Al-Fe-N System. *Met. Trans. A* **1992**, *23*, 3141–3149. [CrossRef]
63. Miettinen, J.; Vassilev, G. Thermodynamic Description of Ternary Fe-B-X Systems. Part 1: Fe-B-Cr. *Arch. Met. Mater.* **2014**, *59*, 601–607. [CrossRef]
64. Seierstein, M. System Al–Fe. COST507. In *Thermochemical Database for Light Metal Alloys*; Ansara, I., Dinsdale, A.T., Rand, M.H., Eds.; European Commission: Luxembourg; Office for Official Publications: Washington, DC, USA, 1998; p. 234.
65. Witusiewicz, V.; Bondar, A.; Hecht, U.; Rex, S.; Velikanova, T. The Al–B–Nb–Ti system. *J. Alloy. Compd.* **2008**, *465*, 64–77. [CrossRef]
66. Maneschi, B.A.S. *Analysis of Nitrides in Boron Steels by Extraction of Nonmetallic Phases*; MET ITAL: New York, NY, USA, 1966.
67. Turkdogan, E.T. Causes and effects of nitride and carbonitride precipitation in HSLA steels in relation to continuous casting. In Proceedings of the 70th Steelmaking Conference Proceedings, Pittsburgh, PA, USA, 29 March–1 April 1987.
68. Sridar, S.; Kumar, R.; Kumar, K.H. Thermodynamic Modelling of Al-B-N System. *Calphad* **2019**, *65*, 291–298. [CrossRef]
69. Wen, H.; Lukas, H.-L. *Thermochemical Database for Light Metal Alloys*; Ansara, I., Dinsdale, A.T., Rand, M.H., Eds.; European Communities: Luxembourg, 1998; Volume 2, System B-N, pp. 65–68.
70. Thermo-Calc Software. CALPHAD Methodology—Thermo-Calc Software. 2023. Available online: https://thermocalc.com/about-us/methodology/the-calphad-methodology/ (accessed on 5 April 2024).
71. Dworkin, A.S.; Sasmor, D.J.; Van Artsdalen, E.R. The Thermodynamics of Boron Nitride; Low-Temperature Heat Capacity and Entropy; Heats of Combustion and Formation. *J. Chem. Phys.* **1954**, *22*, 837–842. [CrossRef]
72. LECO Corporation. LECO. Available online: https://de.leco.com/ (accessed on 7 February 2025).

73. Law, C.A. *Treatise of Petroleum Geology/Handbook of Petroleum Geology: Exploring for Oil and Gas Traps. Chapter 6: Evaluating Source Rocks*; AAPG Special Volumes; AAPG: Tulsa, OK, USA, 1999.
74. Lambda WDS Spectrometers | EDAX. Available online: https://www.edax.com/products/wds/lambda-wds-analysis-system. (accessed on 8 August 2024).
75. Schacht, E.; Richter, J. Erfahrungen mit Ätzmitteln zum Nachweis der ehemaligen Austenitkorngrenzen in Stählen / Experiences with Etching Reagents to Show Former Austenite Grain Boundaries in Steels. *Pr. Met.* **1998**, *35*, 384–395. [CrossRef]
76. Lückl, M. Kinetics Simulation of MnS Precipitation in Electrical Steel. *Steel Res. Int.* **2016**, *87*, 271–275. [CrossRef]
77. Wang, Y.-N.; Bao, Y.-P.; Wang, M.; Zhang, L.-C.; Chen, Y.-N. Basic Research on Precipitation and Control of BN Inclusions in Steel. *Met. Mater. Trans. B* **2013**, *44*, 1144–1154. [CrossRef]
78. Schneider, A.; Stallybrass, C.; Konrad, J.; Kulgemeyer, A.; Meuser, H.; Meimeth, S. Formation of primary TiN precipitates during solidification of microalloyed steels—Scheil versus DICTRA simulations. *Int. J. Mater. Res.* **2008**, *99*, 674–679. [CrossRef]

metals

MDPI

*Article*

# Crystal Plasticity Finite Element Simulation of Tensile Fracture of 316L Stainless Steel Produced by Selective Laser Melting

**Guowei Zeng [1], Ziyang Huang [1], Bei Deng [1] and Rui Ge [2,*]**

[1] Hubei Province Key Laboratory of Systems Science in Metallurgical Process, Wuhan University of Science and Technology, Wuhan 430081, China; zengguowei@wust.edu.cn (G.Z.); ziyanghuang113@outlook.com (Z.H.); dengbeiwust@163.com (B.D.)

[2] State Key Laboratory of Advanced Refractory Materials (Wuhan), Wuhan University of Science and Technology, Wuhan 430081, China

* Correspondence: gerui@wust.edu.cn

**Abstract:** Selective Laser Melting (SLM) of 316L stainless steel exhibits great potential prospects for engineering applications due to its high strength, high forming freedom, and low material waste. However, due to the unique processing technology of additive manufacturing, challenges related to the microstructure and differences in the mechanical properties of the formed parts are inevitable. To investigate the influence of building direction and grain boundary strength on the fracture parameters of SLM 316L stainless steel, electron backscatter diffraction (EBSD) experiments were conducted to characterize the microstructure of SLM 316L stainless-steel specimens. A representative volume element (RVE) model reflecting the microstructure of SLM 316L stainless steel was established based on a combination of the crystal plastic finite element method (CPFEM) and UMAT subroutine technology. The crystal plasticity parameters were determined by comparing the results of tensile tests. Cohesive elements were employed and inserted at the grain boundaries of the polycrystalline RVE to simulate the intergranular fracture behavior of SLM 316L stainless steel under uniaxial tensile loading. The damage and fracture mechanisms of the material at the microscale were analyzed. The simulated tensile stress–strain curves were in good agreement with the experimental results; hence, the combined CPFEM model is suitable for characterizing the mechanical response and fracture behavior of the SLM 316L stainless steel. The results revealed that cracks initiate at stress concentration sites and propagate along grain boundaries with increasing external load, ultimately leading to rupture. Additionally, the building direction influences the location of microcracks and their propagation significantly.

**Keywords:** additive manufacturing; 316L stainless steel; microstructure; fracture behavior; representative volume element

## 1. Introduction

Additive manufacturing presents promising development prospects owing to its distinct advantages: the rapid fabrication of complex structural components, seamless integration of design and manufacturing processes, customization capabilities for individual components, and precise control over localized component properties. Among additive manufacturing technologies, Selective Laser Melting (SLM) stands out for its ability to produce components with superior precision and quality. SLM 316L stainless steel not only exhibits high strength, exceptional forming freedom, and minimized raw material waste, but also demonstrates excellent toughness and superior corrosion resistance. While maintaining mechanical performance comparable to conventionally produced alternatives, it offers significantly reduced costs, thereby facilitating its widespread adoption across aerospace, shipbuilding, medical device manufacturing, and various other industrial sectors [1–3].

In additive manufacturing processes, microstructural variations emerge in components compared to traditional castings due to process parameters, operational procedures, and other influencing factors. These variations manifest as distinctive grain orientations, heterogeneous grain sizes, and defects such as porosity and inclusions. Such microstructural differences directly influence the mechanical behavior of components and represent the primary constraints limiting the broader implementation of additive manufacturing technologies. Consequently, extensive research efforts have focused on elucidating the relationships between process parameters, resultant microstructural characteristics, and mechanical properties through systematic experimental investigations. Song et al. [4] evaluated five distinct scanning strategies to establish correlations between the microstructural features, crystallographic morphology, and mechanical properties of SLM 316L stainless steel. Their findings revealed that scanning strategies exert significant influence on microstructural development, grain growth behavior, grain size distribution, and consequent mechanical properties. The implementation of specific scanning methodologies promotes the formation of equiaxed grain structures, thereby enhancing both tensile strength and ductility in SLM 316L stainless-steel components. Jaskari et al. [5] conducted comprehensive analyses on the microstructure, defect characteristics, and mechanical properties of SLM 316L stainless-steel specimens fabricated using varying volume energy densities. Their research demonstrated that low-energy-density processing conditions generate increased porosity and unfused powder defects, resulting in diminished elongation properties. However, these detrimental phenomena can be effectively mitigated by employing higher-energy-density parameters during fabrication. The research by Zhai et al. [6] demonstrated that incorporating titanium alloy as an additive element during stainless-steel processing refines grain structure (reducing average grain size from 16.7 μm to 0.8 μm) without inducing the formation of detrimental intermediate phases at the grain boundaries in SLM 316L stainless-steel specimens. This microstructural refinement concurrently produces significant enhancements in ultimate tensile strength. Piazza et al. [7] investigated the influences of component geometry and laser power parameters on porosity development and melt pool characteristics in SLM 316L stainless-steel specimens. In complementary research, Fri et al. [8] performed detailed metallographic examinations and heat treatment studies on SLM 316L samples at various temperatures (650 °C, 800 °C, and 1050 °C), characterizing the material's microstructural evolution. Their results confirmed that SLM-produced 316L stainless steel exhibits excellent high-temperature stability that is directly attributable to its distinctive microstructural features. To further elucidate process–microstructure–property relationships in additive manufacturing, Zinovieva et al. [9] examined the effects of scanning patterns on microstructural development and mechanical behavior in SLM 316L stainless steel. Comparing unidirectional and bidirectional scanning strategies, they observed that bidirectional scanning produced microstructures with more pronounced Gaussian and cube crystallographic texture distributions. In contrast, unidirectional scanning resulted in coarser grain structures, more prominent texture patterns, and anisotropic properties. Researchers conducted experimental studies using uniaxial tensile testing [10] and ultrasonic wave velocity methods [11] to obtain the elastic constants of 316L steel under different scanning strategies. Due to the different degrees of shear and tensile forces affecting the interfaces between deposited layers in different printing directions, additively manufactured metals also exhibit significant anisotropic behavior [12–14]. Currently, the reasons for the orthotropic nature of additively manufactured 316L are complex and deserve further investigation.

It is evident that the majority of scholarly research has concentrated on the influence of microstructure on the tensile properties of SLM 316L stainless steel, with comparatively limited investigation into fracture and fatigue behaviors. Suryawanshi et al. [15] conducted a comprehensive evaluation of tensile properties, fracture characteristics, and fatigue crack propagation in SLM 316L stainless steel, benchmarking against conventionally manufactured (CM) austenitic 316L stainless steel. Their experimental findings revealed that SLM 316L stainless steel possesses yield strength compared to its CM counterpart, though at the expense of diminished ductility. Byun et al. [16] employed scanning electron microscopy coupled with electron backscatter diffraction (EBSD) techniques to perform in situ tensile testing on SLM 316L stainless-steel specimens. The EBSD analysis demonstrated that complex microstructural features—encompassing grain morphology, grain orientation differentials, and porosity—exert profound influence on tensile deformation mechanisms and fracture processes. Solberg et al. [17] investigated fatigue behavior in SLM 316L stainless steel through the dual perspectives of internal porosity and surface roughness. Their research established that surface roughness predominantly governs fatigue microcrack initiation under high-cycle fatigue conditions, whereas internal porosity emerges as the principal driver of microcrack formation in low-cycle fatigue regimes.

Experimental methodologies are valuable and constrained by significant drawbacks, including high costs, extensive time requirements, and numerous confounding variables. In contrast, finite element methods are computationally efficient and cost-effective. Contemporary research predominantly utilizes the crystal plastic finite element method (CPFEM) in conjunction with representative volume elements (RVEs) to simulate the mechanical behavior of SLM 316L stainless steel. Ahmadi et al. [18] developed an advanced simulation framework incorporating both grain structure and melt pool characteristics to examine how various specialized microstructures influence the mechanical properties of SLM 316L stainless steel. They integrated cohesive zone elements (CZMs) into their model to simulate mechanical behavior, though notably, their approach did not account for the actual microstructural configurations of test specimens. Cai et al. [19] pioneered an integrated approach combining CPFEM and CZM to investigate the effects of microstructural morphology, grain dimensions, crystallographic orientation, and temperature on the mechanical properties of TWIP steel. Their research conclusively demonstrated that grain boundaries exhibit greater brittleness than grain interiors, resulting in preferential microcrack nucleation at the interfaces between grains. Furthermore, building direction has been shown to significantly impact the mechanical properties of SLM 316L stainless steel [20,21]. This anisotropic behavior stems primarily from preferential grain orientation in the building direction, a critical factor that demands explicit consideration in numerical simulations. In recent years, Nazari-Onlaghi et al. [22] have used various crystal plasticity finite element simulations to model the deformation process of alloys, and Wang et al. [23] inferred that inhomogeneous grain distribution would lead to plastic anisotropy in the stress levels of tensile and compressive behaviors. Therefore, the crystal plasticity finite element method encompasses fundamental aspects of materials science, including lattice structures, dislocation theory, and stress–strain relationships. The application of this method to investigate intragranular microstructure and mechanical behavior provides direct insights into deformation mechanisms under external loading conditions. Consequently, employing crystal plasticity finite element analysis to examine how microstructure and building direction influence fracture behavior in SLM 316L stainless steel contributes to advancing our understanding of damage and failure mechanisms in additively manufactured metallic materials.

This research aims to establish CPFE models and calibrate material parameters based on data obtained from uniaxial tensile experiments and EBSD characterization. Combining CZMs, the proposed CPFE models accurately capture the actual microstructural features of the material, enable the simulation of fracture behavior, and facilitate in-depth analysis of how microstructural characteristics influence fracture mechanisms in SLM 316L stainless steel.

Section 2 presents a brief description of the specimen fabrication of the SLM 316L material and the test methodology, including the methodology of the printing process, the EBSD experiment, and the uniaxial tensile test at room temperature. Section 3 presents and analyzes the uniaxial mechanical properties and microstructure characteristics (EBSD). Section 4 develops a computational framework, integrating crystal plasticity theory with cohesive zone (CZM) element formulations, and outlines the numerical model theory, construct, simulation conditions, substructure model technology, and the results of both the RVE and compact tension tests. Finally, Section 5 summarizes the conclusions and future research directions.

## 2. Specimen Preparation and Experimental Method

SLM316L stainless steel is an austenitic stainless steel with low carbon content, exhibiting excellent corrosion resistance and weldability. The powder size distribution for 316L is determined according to the previous work [15]. The diameter of particles is between 10 μm and 45 μm. Its chemical composition by mass fraction is shown in Table 1.

**Table 1.** Chemical composition of SLM316L stainless steel (weight percentage %).

| Type of Element | Cr | Ni | Mn | Mo | C | Si | Fe |
|---|---|---|---|---|---|---|---|
| Content proportions | 16–18 | 10–14 | 2 | 2–3 | 0.03 | 1 | Remainder |

The SLM316L stainless-steel specimens used in this study were custom-fabricated via the Shenzhen Speed Platform using an SLM280 3D printer manufactured by Sino-Swiss Metals (ZRapid Tech, Suzhou, China). Prior to the experiment, the specimens were treated to ensure good surface conditions and to confirm that they were intact without obvious notches. The dimensional parameters of the specimens are shown in Figure 1a. There are many printing parameters that affect the forming of SLM technology. After a series of trials, the SLM printing parameters used in this study were determined (laser power: 225 W; scanning speed: 1000 mm/s; hatch spacing: 90 μm; powder layer thickness: 30 μm; laser beam diameter: 70 μm). The scanning strategy is illustrated in Figure 1b. Before the experiment, the oxygen concentration in the forming chamber was reduced to below 1000 ppm; 316L stainless steel was used as the base plate, and argon gas was employed as the protective gas. During printing, the layers were rotated by 67° successively. Through layer-by-layer accumulation, the SLM 316L specimens were successfully fabricated according to the two-dimensional data imported into the software before testing. The formed specimen is shown in Figure 1c.

Building upon the work of Suryawanshi et al. [15] and Wang et al. [23], this study conducted room-temperature quasi-static tensile experiments on SLM316L stainless steel in accordance with the previous work [15] at a strain rate of 0.00033 $s^{-1}$ (with a crosshead speed of 0.11 mm/min). Uniaxial tensile tests were conducted at room temperature using an MTS universal material testing machine (Eden Prairie, MN, USA). The tensile direction of the specimens was along the *X*-axis (TD direction), and the experimental data were automatically recorded by a computer.

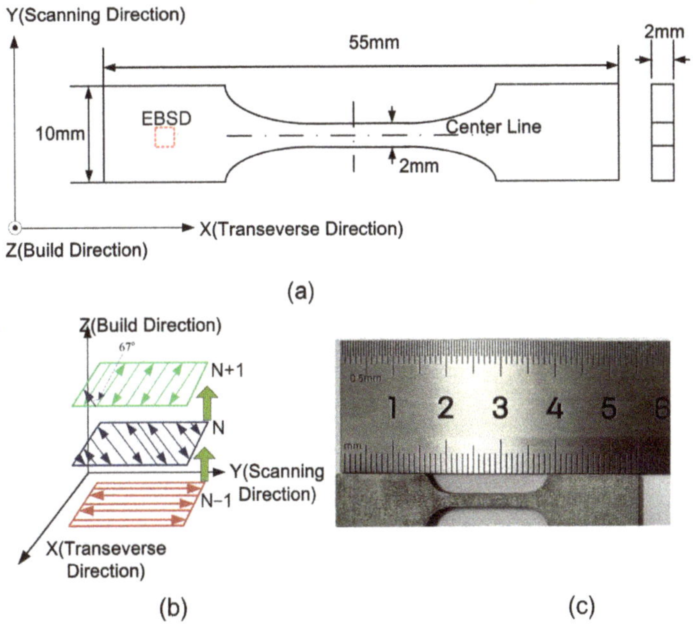


**Figure 1.** Specimen fabrication: (**a**) schematic diagram of specimen dimensions; (**b**) SLM scanning strategy; (**c**) fabricated specimen.

Electron backscatter diffraction (EBSD) has emerged as a critical microstructural characterization methodology in materials science. The successful implementation of this technique necessitates rigorous specimen preparation protocols to achieve both compliance with analytical requirements and optimal data quality [24,25]. The preparation process typically involves sequential mechanical polishing, followed by final chemo-mechanical treatment to obtain ultra-smooth surfaces with minimal topographical variations. Moreover, the meticulous execution of these preparatory stages proves essential for eliminating residual surface damage that could otherwise generate crystallographic artifacts during pattern acquisition. First, a wire-cutting machine was used to cut the complete specimens to obtain samples of appropriate size for the experiment. Then, to ensure that the cut specimen surfaces were flat and smooth, sandpapers of different grit sizes were used to repeatedly perform coarse and fine grinding on the surfaces to be examined. Finally, to ensure the quality of specimen preparation, an electrolytic polishing machine was used for electrolytic polishing treatment. Using a trial-and-error method, parameters of 3 V voltage and 15 A current were set for a 45 s processing treatment. Microstructural observations were conducted using an Apreo S HiVac field-emission scanning electron microscope (FEI company, Hillsborough, OR, USA).

## 3. Experimental Results and Discussion

### 3.1. Microstructure

Figure 2 presents EBSD images of the X-Y plane of the test sample. Figure 2a displays the EBSD inverse pole figure (IPF) of the SLM316L stainless steel. The image clearly demonstrates that the grain morphology of SLM316L stainless steel differs markedly from the coarse grains typically observed in traditionally cast stainless steel. While traditionally cast stainless steel predominantly exhibits regular block-shaped grains, SLM316L stainless steel features elongated grains aligned along the TD direction, frequently spanning multiple melt pools [26]. The diverse colors in the figure represent varying grain orientations, indicating that grain orientations within the X-Y plane do not follow any apparent distribution pattern. However, in some large grain, it can be observed that there are still some continuous trends, and they are distributed along the same direction holistically. It can be observed that although the orientations of the grains are not entirely the same, there are still some continuous trends, and overall, they are distributed along the same direction. Figure 2b illustrates the EBSD grain boundary quality map of the SLM316L stainless steel, where solid lines denote high-angle grain boundaries (>10°) and dashed lines indicate low-angle grain boundaries (2–10°). The analysis of this plane's EBSD image reveals a uniform distribution of both high-angle and low-angle grain boundaries, with high-angle boundaries slightly predominating. Notably, low-angle boundaries typically correlate with high-density dislocations within the material structure.

(a)

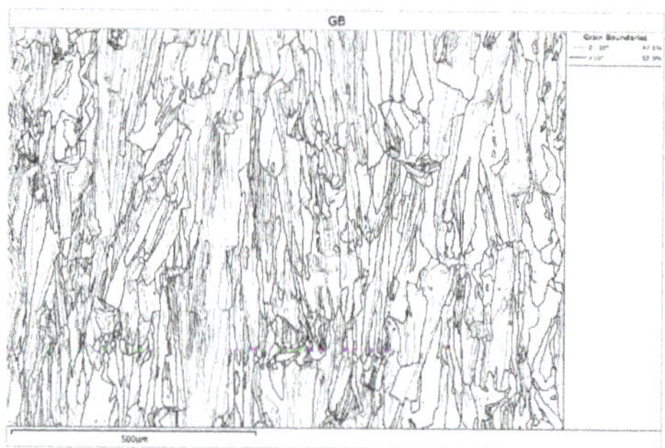


(b)

**Figure 2.** EBSD images of the fabricated sample: (**a**) EBSD inverse pole figure (IPF); (**b**) EBSD grain boundary quality map.

With the entire region serving as the analytical focus, comprehensive grain size distribution data were extracted and statistically analyzed. Figure 3 illustrates the grain size distribution of the fabricated sample. The average grain size throughout the dataset is 26.8 μm, which is considerably smaller than traditionally cast samples. Characteristically, the grain size distribution in the additively manufactured sample is dominated by fine grains (<50 μm), which constitute over 80% of the total population, although larger grains (>50 μm) are also present. Significantly, these larger grains typically function as the primary facilitators of slip occurrence in the material.

### 3.2. Tensile Property

Figure 4a illustrates the measured engineering stress–strain curves, juxtaposed with results from traditionally cast 316L stainless steel (Figure 4b) obtained by Suryawanshi et al. [15]. As can be seen from Figure 4, compared with the traditionally cast 316L specimen, SLM316L stainless-steel specimens demonstrate markedly different mechanical behavior compared to their traditionally cast counterparts. During the elastic regime, stress values escalate rapidly with increasing applied load, approaching the material's elastic strength while maintaining a robust linear stress–strain relationship. Following elastic deformation, the specimen enters an extended plastic hardening phase characterized by a significantly reduced but steady stress growth rate. At elevated stress levels, localized reduction in the specimen's cross-sectional area (necking) occurs, ultimately leading to fracture at a specific location.

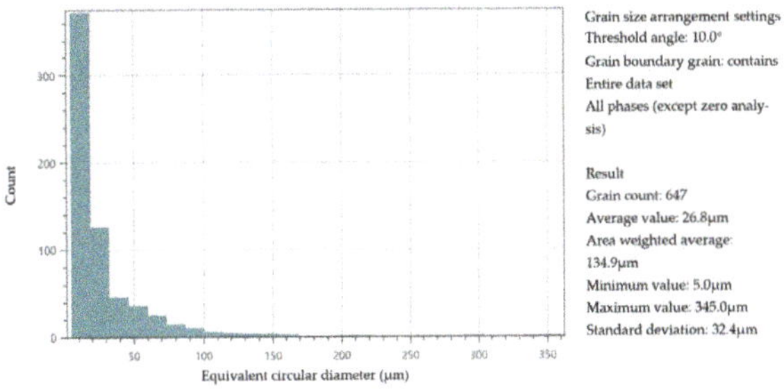


**Figure 3.** Grain size distribution along the X-Y plane of the fabricated sample.

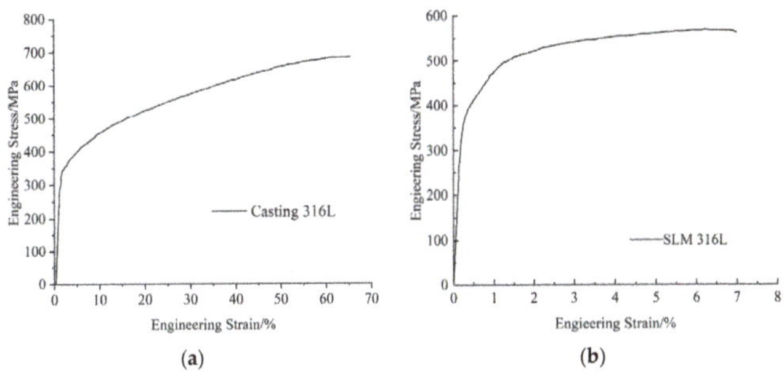


**Figure 4.** Engineering stress–strain curves: (**a**) cast 316L stainless steel [15]; (**b**) SLM316L stainless steel.

Table 2 summarizes the tensile properties of both the SLM316L specimens and reference traditionally cast 316L specimens. The SLM316L stainless steel exhibits a yield strength of 362 MPa, tensile strength of 556 MPa, and elongation of 7.5%, whereas the traditionally cast 316L stainless steel demonstrates a yield strength of 306 MPa, tensile strength of 685 MPa, and remarkably higher elongation of 65%. These comparative data reveal that while SLM316L stainless steel possesses superior yield strength, it underperforms in terms of tensile strength and ductility relative to its traditionally cast counterpart.

**Table 2.** Comparative tensile properties of SLM316L stainless steel and traditionally cast 316L stainless steel.

| Sample | Yield Strength/MPa | Tensile Strength/MPa | Elongation/% |
|---|---|---|---|
| SLM316L | 362 | 556 | 7.5 |
| Rolled316L | 306 | 685 | 65 |

## 4. Finite Element Simulation and Verification

### *4.1. Crystal Plasticity Model*

#### 4.1.1. Kinematics

In this paper, based on Huang's work, a phenomenological rate-dependent plastic constitutive model of the crystal is used to simulate the mechanical response inside the grain. Hill presented a mathematical description of crystal geometry and kinematics in the theory of crystal plasticity. Crystal grains usually undergo lattice rotation and elastic deformation under external loads. If the single crystal has no elastic deformation at all and causes slippage, the gradient of total deformation F is given by

$$F = \mathbf{F^e} \cdot \mathbf{F^p} \tag{1}$$

where gradient elastic deformation $\mathbf{F^e}$ is mainly caused by rigid rotation, and $\mathbf{F^p}$ is plastic deformation gradient caused by shear slip of the crystal along the sliding direction. The plastic deformation gradient satisfies the following equation:

$$\mathbf{L^p} = \sum_{a=1}^{n} \dot{\gamma}^{(\alpha)} \mathbf{s}^{(a)} \mathbf{m}^{(a)} \tag{2}$$

where $\mathbf{s}^{(a)}$ and $\mathbf{m}^{(a)}$ are the slip direction in the reference coordinated system and the sliding surface normal direction.

4.1.2. Constitutive Equations

The stretching rate of the lattice and the Jouman rate of Cauchy stress has a relationship:

$$\overset{\nabla}{\sigma^*} + \sigma(I : D^*) = L : D^* \tag{3}$$

where I and L are the unit tensor and the elastic modulus tensor.

During the slip process, the slip process satisfies SCHMID's law:

$$\tau^{(\alpha)} = m^{*(\alpha)} . \frac{\rho_0}{\rho} \sigma \cdot s^{*(\alpha)} \tag{4}$$

where $\rho_0$ is a reference configuration material density and $\rho$ is the density of the current configuration.

4.1.3. Hardening Laws

According to Schmid's law, each slip system's slip rates of rate-dependent crystalline materials are defined as

$$\dot{\gamma}^{(\alpha)} = \dot{\gamma}_0^{(\alpha)} \mathrm{sgn}(\tau^{(\alpha)} / g^{(\alpha)}) \left| \tau^{(\alpha)} / g^{(\alpha)} \right|^n \tag{5}$$

n is the rate sensitivity index. Here, $g^{(\alpha)}$ is denoted by

$$\dot{g} = \sum_{\beta=1}^{N} h_{\alpha\beta} \dot{\gamma}^{\beta} \tag{6}$$

where $h_{\alpha\beta}$ is slip-enhanced modulus. When $\alpha = \beta$, the hardening mode is self-hardening, and when $\alpha \neq \beta$, the hardening mode is latent-hardening. Self-hardening is adopted as

$$h_{\alpha\alpha} = (h) = h_0 sech^2 \left| \frac{h_0 \gamma}{\tau_s - \tau_0} \right| \tag{7}$$

where $h_0$ represents the initial hardening modulus, $\tau_0$ denotes the material's initial yield stress, and $\tau_s$ signifies the critical shear stress required for plastic flow to occur. The term $\gamma$ corresponds to the sum of accumulated shear strain across all slip systems.

The relationship between the self-hardening modulus and latent-hardening modulus is expressed as

$$h_{\alpha\beta} = qh(\gamma)(\alpha \neq \beta) \tag{8}$$

Here, q is a constant that characterizes the proportional relationship between self-hardening and latent-hardening mechanisms.

In 1991, Bassani and Wu introduced a distinctive expression to describe the hardening modulus of crystalline materials. This formulation encompasses the three characteristic stages of crystal material hardening:

$$h_{\alpha\alpha} = \left\{ (h_0 - h_s) \sec h^2 \left[ \frac{(h_0 - h_s)\gamma^{(\alpha)}}{\tau_s - \tau_0} \right] + h_s \right\} G\left(\gamma^{(\beta)}; \beta \neq \alpha\right) \tag{9}$$

$$h_{\alpha\beta} = q h_{\alpha\alpha} (\beta \neq \alpha) \tag{10}$$

The parameter $h_s$ introduced in this model represents the hardening modulus during the initial stage of easy slip processes in the hardening phase. The function G pertains to interaction hardening and is defined as

$$G\left(\gamma^{(\alpha)}; \beta \neq \alpha\right) = 1 + \sum_{\beta \neq \alpha} f_{\alpha\beta} tanh\left(\frac{\gamma^{(\alpha)}}{\gamma_0}\right) \tag{11}$$

As one of the earliest and most widely implemented crystal plasticity models, this exponential formulation has been extensively utilized in research and subsequently enhanced by numerous investigators. The present study likewise employs this crystal plasticity model as a primary research tool. The aforementioned crystal plasticity formulation is implemented through Huang's UMAT subroutine, which assigns the specific crystal plasticity parameters.

#### 4.1.4. Cohesive Zone Model

The cohesive zone model integrated within Abaqus software 6.14 represents a powerful analytical tool for investigating fracture behavior and has gained widespread adoption in research examining failure mechanisms, particularly delamination in composite materials. In this section, we implement a bilinear cohesive zone model utilizing zero-thickness cohesive elements to simulate both the initiation and propagation of cracks along grain boundaries during tensile loading. The theoretical foundations and physical principles governing these elements are comprehensively documented in the Abaqus manual and thus will not be reiterated here.

Within the Abaqus environment, users can define the maximum threshold value for the damage variable to determine the fate of cohesive elements upon failure. The default setting of 1.0 dictates that when the damage variable reaches this value, the software automatically removes failed cohesive elements from the model, effectively completing the simulation of material fracture processes. For analytical purposes, the SDEG variable, which can be selected in the output settings analysis step, provides real-time tracking of the damage variable, enabling researchers to precisely monitor both the spatial distribution and temporal progression of damage accumulation throughout the loading sequence.

### *4.2. Finite Element Model of Representative Volume Element*

The representative volume element was constructed using the Voronoi tessellation method [27,28], with Neper 4.7.0 [29–31] employed to generate a Voronoi structure comprising 50 distinct grains. Based on experimentally determined grain size measurements, we developed a structural model measuring 0.2 mm in length and 0.12 mm in width. As shown in Figure 5a, each uniquely colored set represents an individual grain, with all 50 single crystal grains collectively constituting the representative volume element of the polycrystalline structure. The random crystallographic orientation data acquired from our EBSD experiments were processed and systematically assigned to each grain set through orientation parameters specified in the UMAT subroutine configuration.

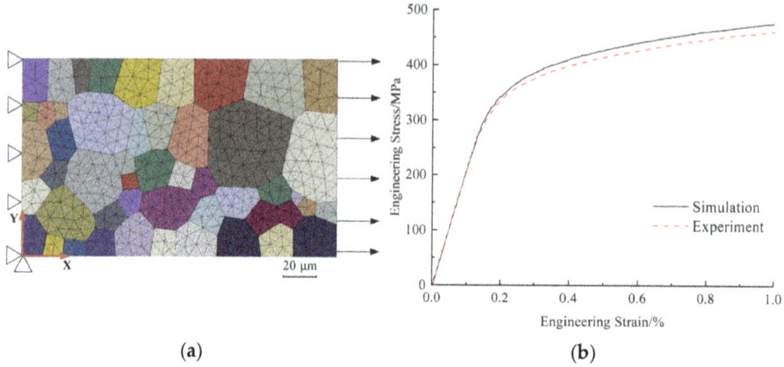


**Figure 5.** Calibration of representative volume element model (RVE): (**a**) meshing and boundary condition of FE model; (**b**) comparison of tensile stress–strain curves obtained from experiment and simulation.

The numerical model utilized CPE3 mesh elements. In order to balance the calculation efficiency and the number of elements, mesh independence studies were conducted. The initial mesh size is set to 10 μm, and the density is gradually increased. After each refinement, the Von Mises stress of the observation point is recorded. Finally, the mesh size is optimally set to 4.8 μm. To eliminate rigid body motion during load application, we implemented strategic boundary constraints. Specifically, the boundary conditions included the following: displacement constraints on the left boundary plane: $U_x = 0$; fixed nodal constraints at the origin located at the lower left corner: $U_x = U_y = 0$. For load application, all nodes on the right boundary plane were coupled and subjected to displacement-controlled tensile strain to simulate uniaxial tension. The loading strain rate is consistent with the experiment.

For the accurate numerical simulation of SLM316L stainless steel, we implemented customized material parameters in the UMAT subroutine [32], with specific parameter sets assigned to individual grains. The elastic modulus of SLM316L stainless steel was precisely determined from tensile test stress–strain data to be $C_{11} = C_{12} = C_{44} = 206$ GPa. The remaining crystal plasticity parameters were established through an optimization process that minimized the mean square deviation between the experimental and simulated stress–strain curves. The complete set of optimized crystal plasticity parameters is presented in Table 3. In the simulation process, SLM316L is considered as having static recrystallization with no phase transformation.

**Table 3.** Crystal plasticity parameters of SLM316L stainless steel.

| Material Parameter | Parameter Value |
|---|---|
| $C_{iiii}$/MPa | 206,000 |
| $C_{iijj}$/MPa | 206,000 |
| $C_{ijij}$/MPa | 206,000 |
| $h_0$/MPa | 171 |
| $\tau_s$/MPa | 214 |
| $\tau_0$/MPa | 156 |
| $n$ | 59 |
| $q$ | 1.11 |
| $\gamma^{(\alpha)}$ | 0.001 |

Figure 5b presents a comparative analysis of the tensile stress–strain response obtained from experimental testing versus finite element simulation. The results demonstrate excellent correlation between experimental and simulated values across the full deformation range, with minor discrepancies remaining well within acceptable tolerance limits for crystal plasticity modeling.

### *4.3. SLM316L Stainless Steel—Fracture Behavior Simulation*

In order to simulate the fracture behavior, the grain boundaries of the RVEs were realized by combining the COH2D4 cohesive elements between adjacent grains in the model shown in Figure 6a. The models in Figures 5 and 6 maintain consistency in boundary condition settings and grain size.

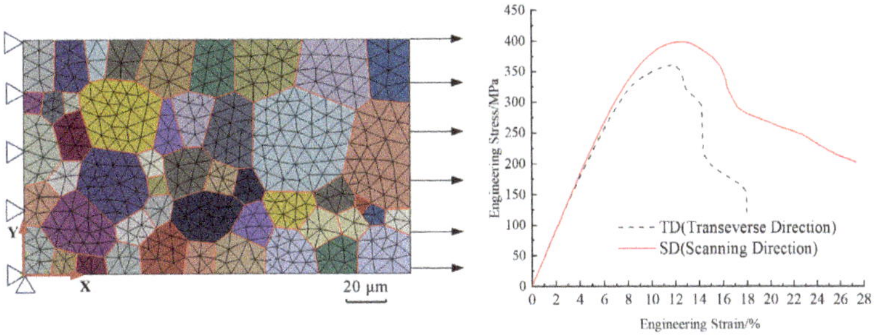


(**a**) (**b**)

**Figure 6.** RVEs combined with cohesive zone elements: (**a**) meshing and boundary conditions of FE model (red lines represent CZM elements); (**b**) comparative stress–strain response curves for RVE models with different grain orientations.

The crystal plasticity parameters employed in this section are consistent with those established in the preceding section. For the cohesive element parameters, it was essential to define the normal stiffness $K_{nn}$, tangential stiffness $K_{ss} = K_{tt}$, maximum normal and tangential traction forces $t_n^0 = t_s^0 = t_t^0$, and energy dissipation rates $G_n$. These specific values are comprehensively detailed in Table 4. The cohesive element thickness was configured at 0.001 mm for all simulations.

**Table 4.** Cohesive element parameter configurations.

| Material Parameter | Numerical Value |
|---|---|
| $K_{nn}$/MPa | 206,000 |
| $K_{ss}$/MPa | 78,000 |
| $K_{tt}$/MPa | 78,000 |
| $t_n^0$/MPa | 550 |
| $t_s^0$/MPa | 550 |
| $t_t^0$/MPa | 550 |
| $G_n$ | 5.5 |

The representative volume elements were configured with two distinct sets of crystallographic orientations. The transverse direction (TD) is basically consistent with the EBSD experimental measurements. Since SLM 316L stainless steel belongs to the face-centered cubic (FCC) structure of austenite, its grain orientation has a high degree of symmetry, and Euler angles within 90° can fully express all grain orientations. A normal distribution probability is used to control the range of Euler angle values, with the specific distribution shown in Table 5. For the scanning direction (SD), the above method is used to randomly generate grain orientations for each grain.

**Table 5.** The distribution probability of Euler angle values.

| Euler Angle | ±7.5° | ±15° | ±22.5° | ±45° |
|---|---|---|---|---|
| $\varphi_1$ | 0.382 | 0.682 | 0.866 | 0.954 |
| $\phi$ | * | * | * | * |
| $\varphi_2$ | 0.382 | 0.682 | 0.866 | 0.954 |

* means the data along the RD (100) direction actually measured by EBSD experiment.

Following the computational analysis of both RVE model sets, the stress–strain curves of different grain directions are presented in Figure 6b. During the elastic deformation stage, the RVE models in the TD direction and the SD direction exhibit the same mechanical response, indicating that grain orientation has almost no influence on the mechanical properties of the material in the elastic deformation stage. However, in the subsequent plastic deformation stage, the yield strength and fracture limit in the SD direction are somewhat higher than those in the TD direction.

Figure 7 illustrates the von Mises stress distribution contour maps for both orientation configurations at the point of computational non-convergence, which corresponds to material failure. The stress contours reveal pronounced heterogeneity in the internal stress distribution across models with different forming directions. While fracture locations vary among different grain boundaries depending on crystallographic orientation, a consistent pattern emerges where fractures in both models initiate at triple junctions—locations where three grain boundaries intersect. A significant disparity in fracture elongation is evident between the two orientation configurations: the SD-direction model demonstrates substantially greater ductility, achieving approximately 27% elongation before failure, whereas the TD-direction model exhibits limited elongation of only about 18%. These computational findings corroborate existing experimental research, which has established that specimens fabricated in the SD direction possess enhanced ductility compared to their TD-direction counterparts. In all cases, fractures initiate at grain boundaries and subsequently propagate along these interfaces until complete material separation occurs.

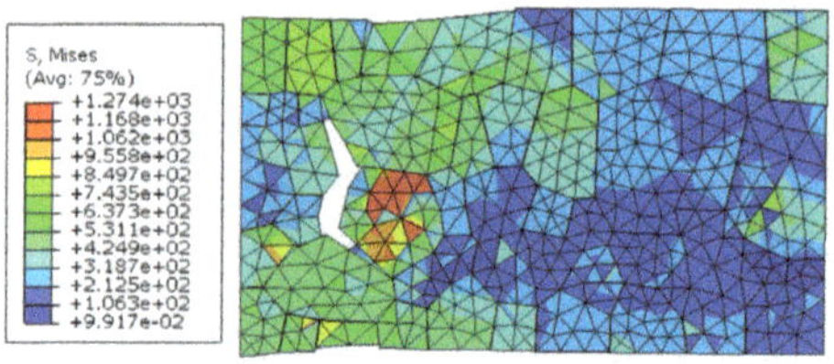


(a)

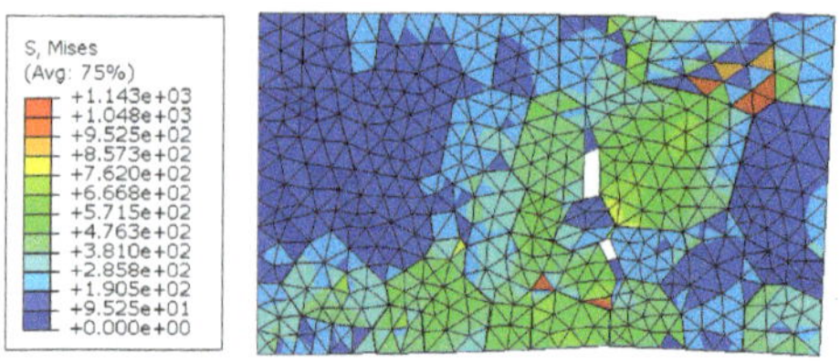


(b)

**Figure 7.** Von Mises stress distribution contour maps of RVE models: (**a**) SD; (**b**) TD.

To elucidate the progressive fracture behavior of the RVE model under uniaxial tensile loading, we extracted and analyzed the damage variable (SDEG) contour maps throughout the crack propagation process, as illustrated in Figure 8. These sequential images capture the complete evolution from crack initiation through propagation to ultimate fracture under sustained uniaxial tension. Five critical stages representing characteristic mechanical behaviors were selected for comprehensive analysis.

At 7.3% tensile strain, the cohesive elements begin to exhibit initial stiffness degradation, signaling the onset of crack nucleation. Although no macroscopically visible cracks or significant cohesive element deformation are apparent at this stage, incipient micro-cracks have formed at the locations highlighted in red, with stiffness degradation quantitatively detectable through numerical analysis. As tensile strain progresses to 9.5%, the cohesive elements manifest substantial stiffness deterioration, with micro-cracks becoming clearly visible at triple junctions where the grain boundaries intersect. These crack initiation sites correspond precisely to regions of stress concentration, conforming to the well-established experimental observation that fracture nucleation preferentially occurs in high-stress regions. Despite this progressive damage to cohesive elements and associated micro-crack formation, the global stress–strain response of the RVE model remains in the strain hardening regime. With continued external loading, cohesive element degradation intensifies and propagates to additional locations, gradually dominating the mechanical response. When tensile strain reaches 14.4%, cohesive elements near a second stress concentration region experience complete degradation (SDEG = 1), indicating total loss of load-bearing capacity. In accordance with the simulation parameters, these failed cohesive elements are removed from the model, resulting in distinct, observable cracks. During subsequent deformation, these primary cracks propagate continuously along grain boundaries with increasing strain, while secondary micro-cracks simultaneously develop throughout other regions of the model. The proliferation of these multiple crack networks progressively diminishes the model's residual strength, initiating a load-shedding process that ultimately culminates in complete material fracture.

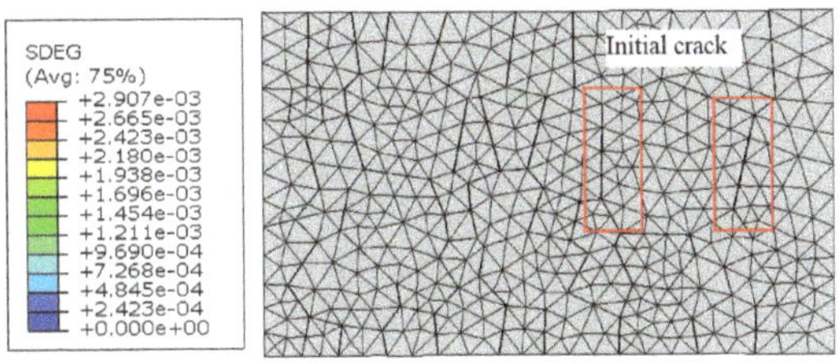


(**a**)

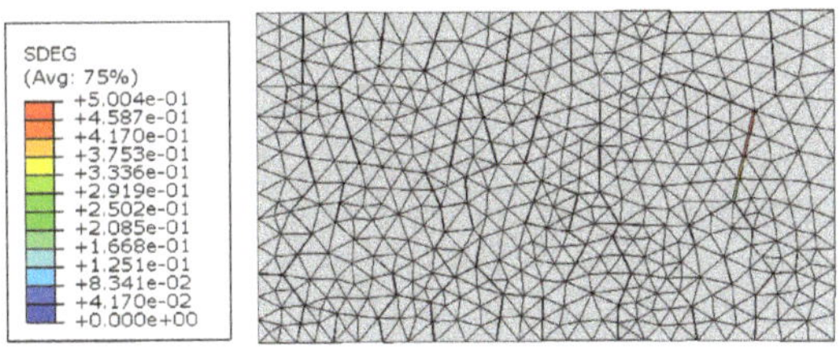


(**b**)

**Figure 8.** *Cont.*

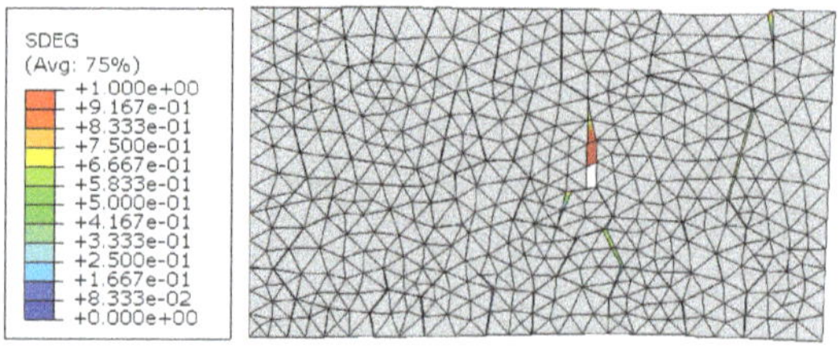


(**c**)

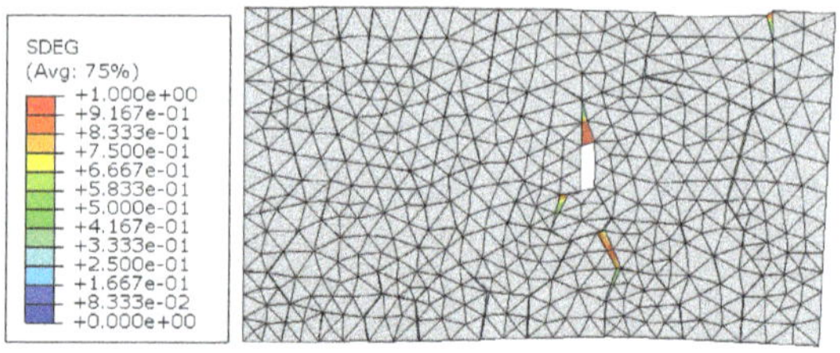


(**d**)

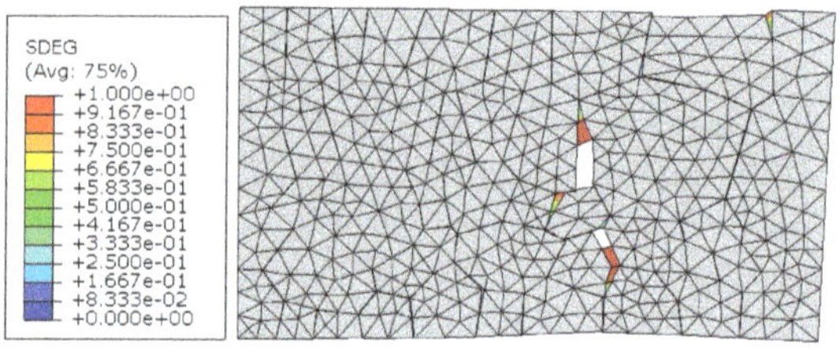


(**e**)

**Figure 8.** Evolution of damage variable (SDEG) distribution in cohesive elements for the TD-direction model: (**a**) $\varepsilon = 7.3\%$; (**b**) $\varepsilon = 9.5\%$; (**c**) $\varepsilon = 14.4\%$; (**d**) $\varepsilon = 16.6\%$; (**e**) $\varepsilon = 17.9\%$.

### 4.4. Influence of Grain Boundary Strength

Grain boundaries, the interfacial regions between adjacent crystalline grains, play a pivotal role in determining both the strength and ductility of metallic materials. The intrinsic strength of these boundaries governs their structural stability and resistance to tensile deformation, thereby significantly influencing the overall mechanical integrity and deformation behavior of the material. In this study, we hypothesized that stainless-steel specimens fabricated using varied process parameters during additive manufacturing would develop fundamentally different grain boundary characteristics and strengths.

To systematically investigate this hypothesis within our computational framework, we maintained the cohesive model parameters established in the preceding section while strategically varying the maximum traction forces of the cohesive elements across three distinct values: 500 MPa, 550 MPa, and 600 MPa. Throughout these simulations, crystal plasticity parameter settings remained constant, while crystallographic orientations were configured to represent both TD (transverse direction) and SD (scanning direction) forming orientations. The resulting stress–strain responses from these six distinct RVE model configurations are illustrated in Figure 9.

The analysis of these results reveals a consistent and significant trend: models with an SD forming direction consistently demonstrate superior mechanical performance, exhibiting both enhanced strength and improved ductility compared to their TD counterparts.

This observation underscores that crystallographic orientation and forming direction exert a more profound influence on overall mechanical behavior than variations in grain boundary strength. Comparative assessment of the stress–strain curves further indicates that incremental increases in maximum traction force (from 500 MPa to 550 MPa to 600 MPa) correlate with progressive improvements in both material strength and ductility.

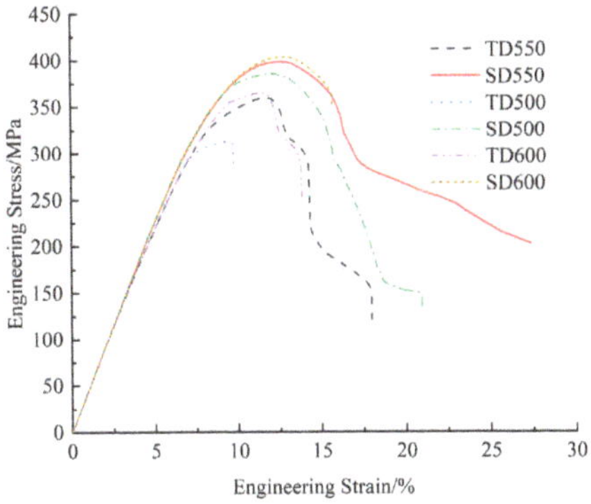


**Figure 9.** Comparative stress–strain responses of RVE models with varying cohesive element parameters.

Notably, however, when the maximum traction force reaches 600 MPa, the material exhibits negligible additional enhancement in mechanical strength. This plateau effect emerges because at such elevated grain boundary strengths, the interfaces no longer constitute the limiting factor in determining fracture strength. Instead, extensive plastic shear strain accumulates within the grain interiors, shifting the dominant failure mechanism from interfacial separation to intragranular deformation. In this regime, the crystal plasticity characteristics become the primary determinant of material strength. During continued loading, widespread crystallographic slip occurs across numerous grains throughout the microstructure, ultimately triggering load redistribution and the onset of material softening behavior.

### 4.5. Fracture Performance Parameter Simulation

The RVE (representative volume element) finite element model provides a sophisticated framework for simulating material mechanical behavior at the mesoscopic scale, effectively characterizing the microscopic mechanical properties of materials under uniaxial tensile loading. In this section, we implement a hierarchical submodeling technique to systematically map the stress–strain responses derived from our RVE models to a macroscopic CT (compact tension) specimen model, thereby facilitating a comprehensive analysis of material mechanical performance at the macroscopic scale.

The material constitutive relationships established through RVE model simulations are strategically incorporated into the CT specimen finite element model depicted in Figure 10. Specifically, the region surrounding the crack tip is designated as the submodel domain, wherein the mechanical response characteristics obtained from the RVE model under uniaxial tension are implemented as the governing elastoplastic constitutive behavior. The remaining portions of the model are assigned the macroscopic mechanical properties of the SLM316L stainless-steel specimen. The CT specimen finite element model is dimensioned with planar measurements of 75 mm × 75 mm and a uniform thickness of 5 mm. Boundary conditions are configured with fixed constraints applied to the coupling point of the lower circular aperture, while the coupling point of the upper circular aperture is subjected to a prescribed displacement of 1 mm. A seam definition is implemented at the crack tip, accompanied by a graduated mesh refinement pattern that progressively increases element density toward the crack tip region. The computational domain is discretized using CPS4 elements (four-node bilinear plane stress quadrilateral elements).

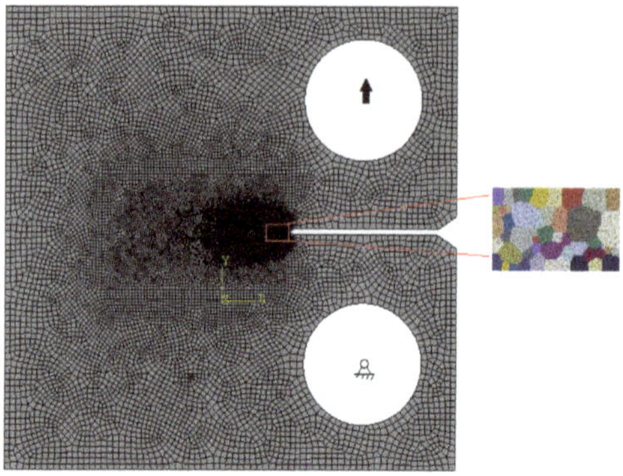

**Figure 10.** Finite element model of the CT specimen.

Figure 11 illustrates six distinct curves representing the J-integral evolution obtained from simulations utilizing the various RVE-derived stress–strain relationships described previously. All simulation results demonstrate consistent behavioral patterns: as the material approaches the fracture threshold, the J-integral values exhibit a pronounced rapid increase, corresponding to significant plastic deformation accumulation in the vicinity of the crack tip. This accelerated growth continues until specific grain boundaries initiate fracture and subsequent load redistribution occurs, at which point the J-integral values transition to a relative plateau. Notably, the J-integral magnitudes associated with models incorporating the SD forming direction consistently surpass those of their TD-direction counterparts across all simulation scenarios. The results further demonstrate that increasing maximum traction force values correlate with enhanced resistance to fracture initiation. When the maximum traction force exceeds a critical threshold, grain boundaries develop substantial resistance to crack propagation, causing the dominant fracture mechanism to transition from interfacial separation to intragranular plastic slip. Collectively, these computational findings establish that specimens fabricated in the SD forming direction exhibit superior comprehensive mechanical performance characteristics, including enhanced yield strength, fracture resistance, elongation capacity, and overall fracture toughness.

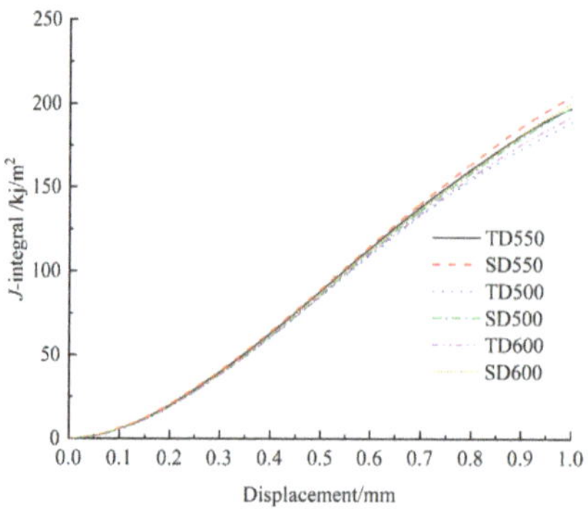


**Figure 11.** J-integral evolution curves obtained from CT specimen finite element simulations.

## 5. Conclusions

Based on crystal plasticity finite element theory, room-temperature uniaxial tensile experiments, materials science experiments, and crystal plasticity finite element simulations were conducted to investigate the microscopic mechanical behavior and fracture properties of SLM316L stainless steel. The main conclusions are as follows:

(1) The RVE model exhibited significant anisotropy at the mesoscopic level. Considering microstructural characteristics, under uniaxial tensile loading, several stress concentration regions appeared, predominantly at triple-grain boundaries. The extracted accumulated plastic shear strain shows a banded distribution, forming a 45° angle with the loading direction. The extracted cumulative plastic shear strain also indicated that grains accumulate varying degrees of plastic shear strain under external loading based on different activated slip systems, showing strong dependence on grain orientation.

(2) Generally, grain boundaries tend to be more brittle than grains. Building upon the established RVE model, cohesive elements were inserted at grain boundaries to simulate fracture behavior, creating an RVE model that reflects forming direction and grain boundary strength. The SD-direction model had an elongation of about 27%, while the TD-direction model had an elongation of only about 18%. The model in the SD direction exhibited delayed fracture compared to the TD direction, and also demonstrated greater elongation at final fracture than the TD-direction model. This is attributed to grain orientations in the SD direction inhibiting slip and dislocation within the grains. Furthermore, the results showed that when the corresponding tensile strain was 9.5%, the cohesive elements had already shown significant stiffness degradation, with microcracks appearing at the triple-grain junctions. When the tensile strain was 14.4%, the cohesive elements failed and macroscopic cracks occurred.

(3) The J-integral of the material simulated by combining the macroscopic CT model and the RVE model during the loading process indicated that the maximum traction force of the cohesive elements also affects the fracture properties of the material. Compared to high grain boundary strength, weaker grain boundary strength caused the cohesive elements in the model to reach the threshold earlier, i.e., the model began to fracture with damage. However, the cohesive elements are no longer the main factor for fracture occurrence when the maximum traction force reaches 600 MPa; they are replaced by the plastic slip within the grains.

In future research, more detailed EBSD and SEM tests will be conducted to replicate and modify the proposed FE model, particularly considering the actual morphology of grain growth in SLM316L stainless steel. A CPFE-CZM model is suggested to enhance the calculation accuracy based on comparisons in the tensile tests. Furthermore, a 3D FE model considering X-Y, Y-Z, and X-Z should be calibrated and validated.

**Author Contributions:** G.Z.: conceptualization, methodology, resources, validation, and supervision; Z.H.: writing—original draft, experimental study; B.D.: methodology, writing—editing and review; R.G.: validation, supervision, and visualization. All authors have read and agreed to the published version of the manuscript.

**Funding:** This research was funded by Key Project of Education Department of Hubei Province [grant D20211108] and Hubei Province Key Research and Development Plan Project [grant 2023BAA019-2].

**Data Availability Statement:** The original contributions presented in this study are included in the article. Further inquiries can be directed to the corresponding author.

**Conflicts of Interest:** The authors declare no conflicts of interest.

## References

1. Wong, K.V.; Hernandez, A. A review of additive manufacturing. *Int. Sch. Res. Not.* **2012**, *2012*, 208760. [CrossRef]
2. Abdulhameed, O.; Al-Ahmari, A.; Ameen, W.; Mian, S.H. Additive manufacturing: Challenges, trends, and applications. *Adv. Mech. Eng.* **2019**, *11*, 1687814018822880. [CrossRef]
3. Frazier, W.E. Metal additive manufacturing: A review. *J. Mater. Eng. Perform.* **2014**, *23*, 1917–1928. [CrossRef]
4. Song, Y.N.; Sun, Q.; Guo, K.; Liu, F.; Ma, J. Effect of scanning strategies on the microstructure and mechanical behavior of 316L stainless steel fabricated by selective laser melting. *Mater. Sci. Eng. A* **2020**, *793*, 139879. [CrossRef]
5. Jaskari, M.; Ghosh, S.; Miettunen, I.; Karjalainen, L.P. Tensile properties and deformation of AISI 316l additively manufactured with various energy densities. *Materials* **2021**, *14*, 5809. [CrossRef]
6. Zhai, W.; Zhou, W.; Nai, S.M.L. Grain refinement of 316L stainless steel through in-situ alloying with Ti in additive manufacturing. *Mater. Sci. Eng. A* **2022**, *840*, 142912. [CrossRef]
7. Piazza, S.; Merrigan, B.; Dowling, D.P.; Monaghan, M.; Lennon, A.; Purcell, P.J. The effects of geometry and laser power on the porosity and melt pool formation in additively manufactured 316L stainless steel. *Int. J. Adv. Manuf. Technol.* **2020**, *111*, 1457–1470. [CrossRef]
8. Fri, K.; Laazizi, A.; Bensada, M.; Diyer, Z.; Bahram, A. Microstructural and heat treatment analysis of 316L elaborated by SLM additive manufacturing process. *Int. J. Adv. Manuf. Technol.* **2023**, *124*, 2289–2297. [CrossRef]
9. Zinovieva, O.; Romanova, V.; Balokhonov, R. Effects of scanning pattern on the grain structure and elastic properties of additively manufactured 316L austenitic stainless steel. *Mater. Sci. Eng. A* **2022**, *832*, 142447. [CrossRef]
10. Zhang, L.; Ma, Q.; Ding, J.; Qu, S.; Fu, J.; Fu, M.W.; Song, X.; Wang, M.Y. Design of elastically isotropic shell lattices from anisotropic constitutive materials for additive manufacturing. *Addit. Manuf.* **2022**, *59*, 103185. [CrossRef]
11. Javidrad, H.R.; Salemi, S. Determination of elastic constants of additive manufactured Inconel 625 specimens using an ultrasonic technique. *Int. J. Adv. Manuf. Tech.* **2020**, *107*, 4597–4607. [CrossRef]
12. Zinovieva, O.; Zinoviev, A.; Patel, M.N.; Molotnikov, A.; Easton, M.A. Modelling grain refinement under additive manufacturing solidification conditions using high performance cellular automata. *Mater. Des.* **2024**, *245*, 113248. [CrossRef]
13. Barkia, B.; Vallet, M.; Tanguy, A.; Auger, T.; Héripré, E. New insights into microstructure evolution and deformation mechanisms in additively manufactured 316L stainless steel. *Mater. Sci. Eng. A* **2025**, *934*, 148327. [CrossRef]
14. Li, J.N.; Gao, D.; Lu, Y.; Hao, Z.P.; Wang, Z.Q. Mechanical properties and microstructure evolution of additive manufactured 316L stainless steel under dynamic loading. *Mater. Sci. Eng. A* **2022**, *855*, 143896. [CrossRef]
15. Suryawanshi, J.; Prashanth, K.G.; Ramamurty, U. Mechanical behavior of selective laser melted 316L stainless steel. *Mater. Sci. Eng. A* **2017**, *696*, 113–121. [CrossRef]
16. Byun, T.S.; Gussev, M.N.; Lach, T.G. Deformation and Fracture Behavior of Additively Manufactured 316L Stainless Steel. *JOM* **2024**, *76*, 362–378. [CrossRef]
17. Solberg, K.; Guan, S.; Razavi, S.M.J.; Welo, T.; Chan, K.C.; Berto, F. Fatigue of additively manufactured 316L stainless steel: The influence of porosity and surface roughness. *Fatigue Fract. Eng. Mater. Struct.* **2019**, *42*, 2043–2052. [CrossRef]
18. Ahmadi, A.; Mirzaeifar, R.; Moghaddam, N.S.; Turabi, A.S.; Karaca, H.E.; Elahinia, M. Effect of manufacturing parameters on mechanical properties of 316L stainless steel parts fabricated by selective laser melting: A computational framework. *Mater. Des.* **2016**, *112*, 328–338. [CrossRef]
19. Cai, W.; Sun, C.; Wang, C.; Gao, Y.; Chen, K. Modelling of the intergranular fracture of TWIP steels working at high temperature by using CZM–CPFE method. *Int. J. Plast.* **2022**, *156*, 103366. [CrossRef]
20. Marattukalam, J.J.; Karlsson, D.; Pacheco, V.; Beran, P.; Wiklund, U.; Jansson, U.; Hjörvarsson, B.; Sahlberg, M. The effect of laser scanning strategies on texture, mechanical properties, and site-specific grain orientation in selective laser melted 316L SS. *Mater. Des.* **2020**, *193*, 108852. [CrossRef]
21. Zhao, C.; Bai, Y.; Zhang, Y.; Wang, H.; Chen, C.; Shi, J. Influence of scanning strategy and building direction on microstructure and corrosion behaviour of selective laser melted 316L stainless steel. *Mater. Des.* **2021**, *209*, 109999. [CrossRef]
22. Nazari-Onlaghi, S.; Li, G.; Vanmeensel, K.; Seefeldt, M. Deformation texture in LPBF 316L studied by different crystal plasticity models. *Mater. Sci. Eng. A* **2025**, *924*, 147769. [CrossRef]
23. Wang, Z.; Jiang, B.; Wu, S.; Liu, W. Anisotropic tension-compression asymmetry in SLM 316L stainless steel. *Int. J. Mech. Sci.* **2023**, *246*, 108139. [CrossRef]
24. Nowell, M.M.; Witt, R.A.; True, B.W. EBSD sample preparation: Techniques, tips, and tricks. *Microsc. Today* **2005**, *13*, 44–49. [CrossRef]
25. Wilkinson, A.J.; Britton, T.B. Strains, planes, and EBSD in materials science. *Mater. Today* **2012**, *15*, 366–376. [CrossRef]
26. Andani, M.T.; Karamooz-Ravari, M.R.; Mirzaeifar, R.; Ni, J. Micromechanics modeling of metallic alloys 3D printed by selective laser melting. *Mater. Des.* **2018**, *137*, 204–213. [CrossRef]

27. Aurenhammer, F.; Klein, R. Voronoi Diagrams. In *Handbook of Computational Geometry*; Sack, J.R., Urrutia, J., Eds.; Elsevier: Amsterdam, The Netherlands, 2000; pp. 201–290.
28. Du, Q.; Faber, V.; Gunzburger, M. Centroidal Voronoi tessellations: Applications and algorithms. *SIAM Rev.* **1999**, *41*, 637–676. [CrossRef]
29. Quey, R.; Dawson, P.R.; Barbe, F. Large-scale 3D random polycrystals for the finite element method: Generation, meshing and remeshing. *Comput. Methods Appl. Mech. Eng.* **2011**, *200*, 1729–1745. [CrossRef]
30. Quey, R.; Renversade, L. Optimal polyhedral description of 3D polycrystals: Method and application to statistical and synchrotron X-ray diffraction data. *Comput. Methods Appl. Mech. Eng.* **2018**, *330*, 308–325. [CrossRef]
31. Quey, R.; Villani, A.; Maurice, C. Nearly uniform sampling of crystal orientations. *J. Appl. Crystallogr.* **2018**, *51*, 1162–1173. [CrossRef]
32. Huang, Y. A User-Material Subroutine Incorporating Single Crystal Plasticity in the ABAQUS Finite Element Program. Ph.D. Thesis, Harvard University, Cambridge, MA, USA, 1991.

*metals*

*Article*

# Verification of Numerical Models of Steel Bar Coverings Using Experimental Tests—Preliminary Study

**Paweł Zabojszcza [1], Krystyna Radoń-Kobus [2] and Paweł Grzegorz Kossakowski [1,*]**

[1] Faculty of Civil Engineering and Architecture, Kielce University of Technology, 25-314 Kielce, Poland; pawelzab@tu.kielce.pl
[2] Faculty of Mechatronics and Mechanical Engineering, Kielce University of Technology, 25-314 Kielce, Poland; kradonkobus@tu.kielce.pl
* Correspondence: kossak@tu.kielce.pl

**Abstract:** In the design of metal bar coverings, the key problem is to correctly determine the numerical model of the analyzed structure. The description of numerical models may differ from the actual, real behavior of the structure. Therefore, there is a need to verify and calibrate them using experimental studies. The aim of this research will be to verify and assess the accuracy of the numerical model of a metal bar roof by conducting experimental studies. A series of repeatable experimental tests will be conducted on the structure model to determine the path of static equilibrium and the form of stability loss of the steel covering. During the test, as the load increases, data will be collected on the displacements of nodes. The displacements of the nodes will be verified using precise triangulation laser sensors and electronic sensors. Based on the results of the tests, conclusions will be drawn regarding the accuracy of the numerical models. Comparison of the results obtained from the numerical models with the experimental data will allow for the identification of possible discrepancies and understanding how the numerical models can be improved. This in turn will contribute to the development of more advanced and more accurate methods for the analysis of metal bar roof structures in the future.

**Keywords:** finite element method; metal bar roof; equilibrium path; stability loss; imperfections

## 1. Introduction

Construction failures resulting in the damage of load-bearing structures made of steel members are one of the most severe not only in terms of material loss but also because of potential life and health loss. The phenomena related to stability may be decisive in the design of steel roofs. Stability theory [1–5] is distinguished in structural mechanics by the need to depart from the stiffness principle, which means that nonlinear equations of at least the second order should be used. Geometrically nonlinear analysis (GNA) allows a complete description of the potential form of stability loss. The mathematical model of nonlinear discrete systems formulated on the basis of the FEM corresponds to a system of nonlinear algebraic equations [6,7]. Methods for solving nonlinear systems of equations are often based on the Newton–Raphson algorithm, in which points lying on the equilibrium path are determined in successive steps. In each step of the algorithm, a sequence of iterations is performed in such a way that at the end of the step, the solution is obtained with the accuracy of the given error. Depending on the step method (control), we can distinguish: load, displacement, or arc parameter control. The basic problem in numerical analysis of nonlinear problems is the occurrence of singular points on the equilibrium path (bifurcation points, limit points, turning points). At these points, standard algorithms for solving systems of linear equations fail. In this paper, we intend to use the constant arc length method [8,9] to determine and evaluate equilibrium paths. The structures analyzed in this paper will be made of class 1 steel members. In some situations, there is a need to track equilibrium positions after exceeding critical load or displacement values (post-critical states). If the critical parameters of the load correspond to qualitatively different equilibrium positions or if qualitatively different post-critical states are possible, then such a critical equilibrium position is called a bifurcation. The case when, after exceeding the state of critical equilibrium, only unstable equilibrium positions occur corresponding to the values of the load parameters lower than the critical ones is called the limit equilibrium position. It happens that a further increase in displacements causes the system to move to another, distant equilibrium position, which may turn out to be stable. This phenomenon is called node snapping. In the case of low steel bar coverings, this form of stability loss is often decisive. Numerical analysis of steel bar coverings is the subject of many works. In [10], the authors describe the influence of ball joint deformations on the inelastic buckling behavior of domes and vaulted structures. The authors present results for single-layer domes and a stiffened barrel vault. In [11], the phenomenon of node snapping of shallow lattice domes is analyzed. The equilibrium paths are presented for different load combinations. In the paper [12], a few shallow domes were produced using a 3D printer and tested using a load cell and proximity laser sensor. The lattices studied here incorporate moment-transmitting joints and clamped boundary conditions at the perimeter. The flexure is the dominant mode of deformation under lateral loading. The complete displacement–load relationship and multiple equilibrium configurations are shown both in simulation and in experiment. In [13], the authors analyze elastic space trusses in the shape of a regular pyramid. Equilibrium paths are derived for each load condition under the hypotheses of linear material law, small or moderate axial deformation in bars, and large nodal displacements. These mechanical systems exhibit a wide variety of post-critical responses and are not exhausted by the classical snapping and bifurcation phenomena. In addition to regular primary and secondary branches, the equilibrium paths may include neutral branches. These branches are particularly difficult to handle by the standard numerical procedures of nonlinear analysis. Despite the efforts to unify nonlinear solution methods, researchers have not yet created an ideal algorithm that could correctly determine the equilibrium paths for all problems. Verification of these methods requires accurate equilibrium equations of some structures. In the case of structures susceptible to large displacement gradients, complicated behaviors after buckling with different critical points are observed. The analysis of equilibrium paths and their features, along with the provision of exact expressions, is described in [14]. Detailed experimental study and numerical simulation of the progressive collapse resistance of single-layer latticed domes are shown in [15,16]. The nonlinear finite element model used to simulate equilibrium paths is presented in [17]. Special interest is focused on the identification of critical points which contribute to the understanding and quantification of stability processes in a nonlinear system. The paper discusses the methods of stochastic and sensibility analyses which are frequently applied to the assessment of the safety and reliability of supporting structural systems. It is pointed out that the failures of the slender structural system caused by stability loss very often causes its decrease in safety and reliability. Therefore, reliable methods of stability analysis are important instruments to ensure the general effectiveness of modern structural systems. The snap-through phenomenon is very important in the stability analyses of steel members with initial imperfection. It is also observed in real engineering structures, as well as during experiments. This issue is studied by the authors of [18] on the example of geometrically nonlinear truss structures. By tracing their equilibrium paths, using an arc length approach within a finite element analysis, snap-through behavior is demonstrated. The snap-through phenomenon of the Mises truss taking into account spatial asymmetric and symmetric buckling with an out-of-plane lateral linear spring is also investigated in [19]. In the paper, the tangent stiffness matrix, for the primary equilibrium trajectory, is obtained in a closed form and eigenvalues of the stiffness matrix are also derived analytically. They are used for the detection, classification, and sequencing of critical points in the primary equilibrium path of the Mises truss. In the article [20], the authors presented an analysis of a low-rise rod dome subjected to wind loading in terms of verifying the type of stability loss and the influence of geometric imperfections on the dimensioning of the structure's bars. The article [21] proposed verification of the stability of a frame rod structure using a probabilistic description. The analysis involved a rod dome for which the standard safety level (reliability index) was compared with the determined reliability index for the designated load of the structure. Verification of critical load-bearing capacity is an extremely important aspect of structural design. The paper [22] focuses on the influence of the node density of a geodesic dome on its critical load-bearing capacity. The authors considered both truss and frame models. Design of structures focuses on verification of load-bearing capacity and verification of displacements assuming static loads. Verification of the dynamic response of the structure to applied loads (including wind load) is often an omitted issue. In the paper [23], the authors attempted to verify the influence of the geometry of bar domes on the results of modal analysis. Two domes were verified: a Schwedler dome and a geodesic dome. Additionally, the domes were verified in terms of stability loss.

In the articles [24,25], thin-walled elements made of cold-formed channel sections were analyzed. The static analysis of the buckling state and after buckling of sections subjected to eccentric compression was presented. Equilibrium paths for different values of eccentricities were determined, and deformation maps of the analyzed sections were presented. The authors compared numerical analyses with experimental studies. In the studies described in [26], the buckling characteristics of spherically segmented toroidal shells subjected to external pressure were determined. The authors tested and evaluated two laboratory models with the same nominal dimensions. The sensitivity of the tested models to imperfections with closely spaced eigenvalues was verified. The first 50 eigenmodes were used for the evaluation. Additionally, the post-buckling behaviour of the models was evaluated, for which a modified Riks algorithm was used. The results indicated that the deviation between the results of the experimental and numerical analyses was within a reasonable range of about 14%.

Numerical problems of controlling the incremental-iterative process during the generation of equilibrium paths are very important. Correct modeling of the structure is an equally important issue. The model of the structure should take into account the real structure as faithfully as possible. The difficulty of this stage can be seen when performing experimental tests and numerical calculations. Initially, we most often verify the description of the ideal structure (we assume perfectly straight bars, supports as non-deformable). In practice, however, we rarely deal with such a situation. In order to bring the numerical model closer to the real structure, it is often necessary to check the flexibility of the supports and imperfections of the bars, which is often not a simple task. A number of scientific works have been devoted to these issues. Nonlinear FEM formulations based on von Kármán kinematics with initial geometric imperfections were considered in [27]. The authors included a reduced-order predictor-corrector modeling method for nonlinear buckling analysis of thin-walled structures with initial geometric imperfections based on Koiter's theory. In this work, the biggest innovation of the introduced method is the implementation of an initial geometric imperfection instead of equivalent loads. This allows the application of the reduced-order Koiter-Newton modeling method to pre- and post-buckling buckling, even for relatively large imperfection amplitudes. In [28] the idea of increasing the buckling resistance by means of modal geometric imperfections is discussed. The authors pointed out that not all imperfections are harmful for buckling problems. A recent design trend called "modal nudging" uses geometric imperfections to change the original structure to equilibrium paths with higher capacity. An automated procedure based on the reduced-order model inspired by Koiter is proposed, enabling fast estimation of equilibrium paths of structures with geometric imperfections according to the first linearized buckling modes. In this method, the selection of the geometric imperfection that pushes the structure to a favorable equilibrium path is performed by means of a fast random search. In such cases, geometric imperfection can play a favorable role. The "Modal nudging" method aims to improve the buckling response of slender elastic structures by introducing a small change in the base geometry, capable of pushing the structure towards higher equilibrium paths, both for unstable and stable equilibrium paths. However, the authors noted that guaranteeing a response that is insensitive to imperfections is not sufficient. In fact, designers should be aware that small random imperfections can push the structure back towards unstable paths. The paper [29] addresses the subject of stability of steel elements with a complex initial pattern of geometric imperfections. This aspect is extremely important for slender structures, which are characterized by many close critical loads and modal actions, leading to possible unstable postcritical paths and imperfection sensitivity. A model of a structure consisting of four rigid bars connected by elastic nodes is analyzed. For which a set of nonlinear algebraic equilibrium equations is developed. The formulas are considered for structures with and without initial geometric imperfections. In order to develop a complex imperfection formula, a linear superposition of buckling modes obtained from a linear eigenvalue problem is used. The nature of bifurcation points is studied, adopting the concept of minimum potential energy. The influence of imperfections on the behavior of the structure is verified based on numerous examples, finding that for specific imperfection patterns, an increase in the amplitude of the initial geometric imperfection can result in an increase in the critical load defining the bifurcation point. Cases in which initial geometric imperfections can play a positive role result in stable behavior after buckling. This phenomenon corresponds to the so-called "modal nudging". Its aim is to improve the buckling response of slender elastic structures by introducing a small perturbation in the initial geometry of the structure. This can result in equilibrium paths with higher load-bearing capacity. A snap through phenomenon caused by the transition from the local to the global minimum of potential energy is also observed. This observation was caused by a specific configuration of the initial geometric imperfections, which in this case played a rather dangerous role. It is crucial that the proposed model structure allows for a complete description of the post-critical behaviour as well as for tracing the impact of complex imperfection configurations in a simple and transparent way.

The task of verifying numerical simulations using experimental tests is a very important and difficult task. First of all, it allows us to detect possible errors and better understand the work of the structure. The authors of the paper tried to take up this task. In this paper, example of the low steel bar covering is analyzed. This structure is characterized by a strongly nonlinear static response. Ignoring nonlinear effects leads to large computational errors. Structural analysis is carried out in accordance with EN 1990:2002 and EN 1993-1-1 [30,31] using Abaqus Learning Edition 2022 software. The main aim of the research is to verify and assess the accuracy of numerical models of steel bar roof by conducting experimental studies. At the beginning a mechanical and chemical properties of the material are calculated. Next a series of repeatable experimental tests are conducted on the structure model to determine the static equilibrium path. During the test, as the load increases, data are collected on the displacements of nodes. The displacements of nodes are verified using precise triangulation laser sensors and electronic sensors. Based on the results of the tests, conclusions will be drawn regarding the accuracy of the numerical models. Comparison of the results obtained from the numerical models with the experimental data will allow for the identification of possible discrepancies and understanding how the numerical models can be improved. This in turn will contribute to the development of more advanced and more accurate methods for the analysis of metal bar roof structures in the future.

The structure of the paper is as follows. The next section is devoted to the detailed description of material parameters. The chemical composition of the tested steel is examined using a Phenom XL scanning electron microscope (Thermo Fisher Scientific, Waltham, MA, USA). Next, we checked mechanical properties of the material: tensile strength, yield strength, Young's modulus. Then, the experimental model of the structure and load are defined. The next section briefly describes nonlinear geometrical analysis by the finite element method. Section 3 presents the obtained results. This paper ends with a discussion of the obtained results and conclusions.

## 2. Materials and Methods

### 2.1. Description of a Mechanical and Chemical Properties of the Material

The surface and chemical composition of the samples were examined using a Phenom XL scanning electron microscope. During the tests, an accelerating voltage of 15 kV and a magnification of ×5000 were used. This allowed for precise determination of the morphology of the samples and confirmation of their chemical composition.

The geometric structure of the surface of the samples was observed before testing the surface wettability with a Leica DCM8 (Leica, St. Gallen, Switzerland) optical microscope in confocal mode. Observations were performed in confocal mode at a magnification of ×20. The measured surface area was 1.22 mm × 1.78 mm. The measurement was performed three times at different locations on the surface.

The Theta Flex optical tensiometer from Attension (Biolin Scientific AB, Espoo, Finland) was used to examine the contact angles of steel using the sessile drop method. The measurement was carried out by precisely placing a 5 μL drop of measuring liquid—distilled water on the surface of the samples. After approximately 1 s from the moment the drop settles on the sample surface, photos are taken using a digital camera. Figure 1 shows the scheme for determining the wetting angle value. The measurements were repeated five times at randomly selected locations on the sample and based on them, the average values of the wetting angle were determined.

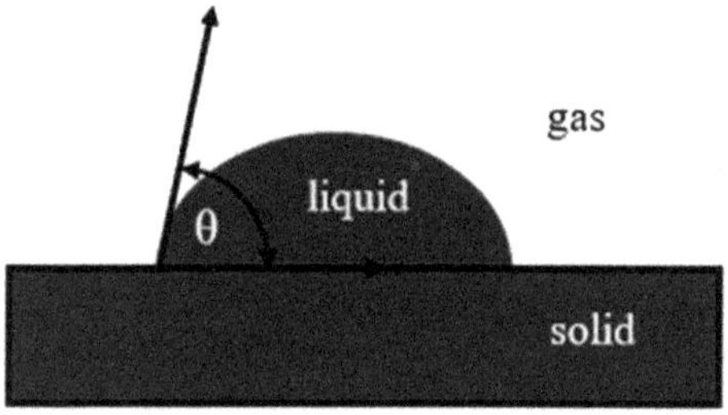


**Figure 1.** Schematic diagram of a liquid droplet applied to the sample surface (hydrophilic surface); θ—equilibrium contact angle (Young's angle).

Figure 2 shows SEM images of steel showing the surface microstructure observed at ×10,000 magnification and the distribution of elements on the tested surface. Table 1 contains the chemical composition of the tested steel determined using a scanning electron microscope. The map of the distribution of elements shows the most intense color for iron. This indicates its largest share on the tested surface. Manganese has a less intense color, which means its content is a much lower percentage. The colors on the map describing silicon, chromium, copper, and nickel are characterized by similar intensity. The lowest intensity and density of color are presented for phosphorus and sulfur, which confirms a very similar percentage content of these elements in the material.

**Table 1.** Chemical composition of steel.

| Steel [% Mass Content] | | | | | | | | |
|---|---|---|---|---|---|---|---|---|
| Fe | C | Mn | Si | Cr | Cu | Ni | P | S |
| 97.26 | 0.4 | 1.32 | 0.35 | 0.22 | 0.18 | 0.16 | 0.06 | 0.05 |

The chemical composition of the tested steel, presented in Table 1, indicates a predominant iron content. The analysis of the chemical composition revealed a higher carbon content than originally assumed. The additions of manganese, silicon, chromium, copper, nickel, and phosphorus are within the assumed limits.

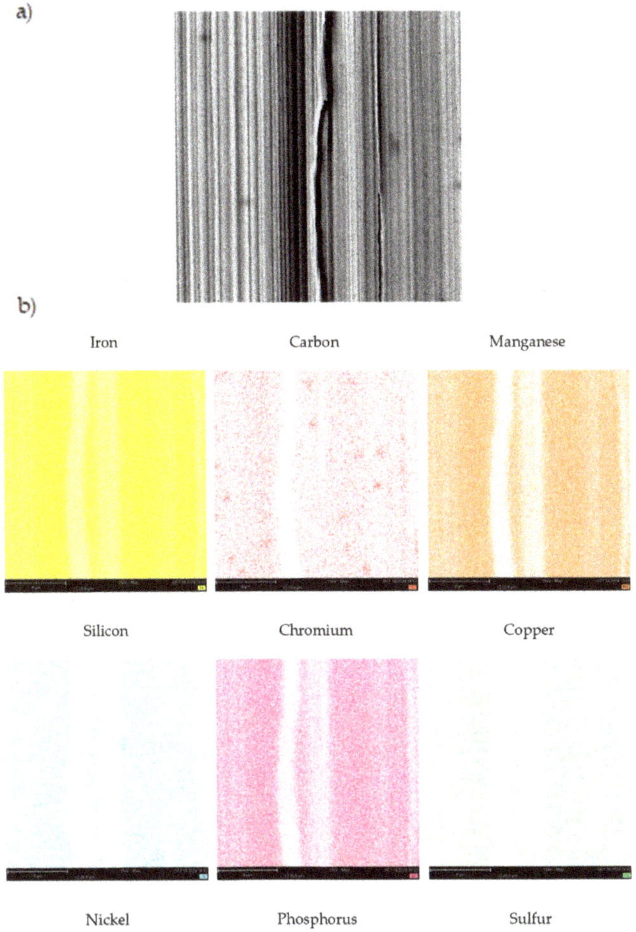


**Figure 2.** SEM image of steel: (**a**) microstructure magnification ×10,000, (**b**) distribution of elements on the surface.

The tested surface was characterized by the following amplitude parameters Sq = 3.2 µm, Sku = 4.34, Ssk = −0.64, and Sa = 2.48 µm. This proves that the tested surface is very developed. The average contact angle was about 41°, which indicates its hydrophilic surface (Figure 3).

The study of the applied materials is an important aspect of the research work. The authors in the work [32] evaluated the effect of surface roughness on the properties of a-C:H:Si coatings obtained using plasma-assisted chemical vapor deposition (PACVD). The substrate was characterized by, among others methods, microscopic observation, thickness measurements, wettability, surface topography, and tribological tests. A scanning electron microscope with energy-dispersive spectroscopy (EDS) was used to examine the coating microstructure, thickness, and chemical content. The authors used a confocal microscope to examine the geometric structure of the surface before and after the tribological tests. The tribological tests used a ball-on-disc sliding configuration in reciprocating motion under dry friction and cutting oil lubrication. The work [33] concerns the study of the effect of DLC coatings doped with tungsten coatings (a-C:H:W) obtained using plasma vapor deposition (PVD) on the properties of 100Cr6 steel. Samples of 100Cr6 steel with and without coating were verified. In the studies, scanning electron microscopy (SEM) was used to observe the surface morphology of coatings and identify the elements in the composition of the coating. The authors assessed the effect of coatings on the tribological properties of lubricated friction pairs. Based on the studies, it was found that the use of DLC a-C:H:W coatings increases the hardness of 100Cr6 steel threefold. The diamond-like coating with a tungsten dopant in friction conditions reduced the coefficient of friction and wear. DLC a-C:H:W coatings lubricated with a cutting fluid improve the mechanical and tribological properties of the sliding surfaces of 100Cr6 steel in friction conditions.

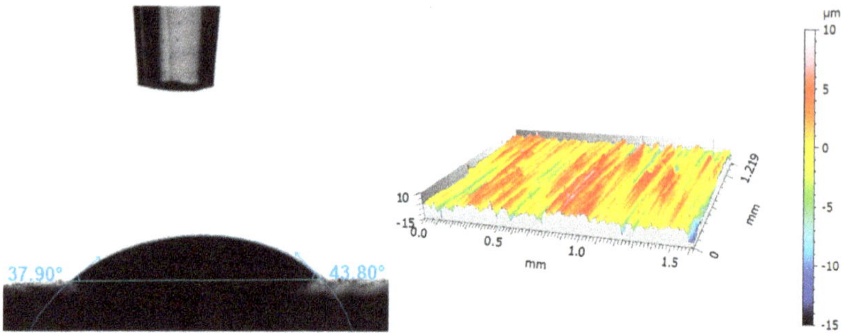


**Figure 3.** Examples of photographs showing demineralized water droplet and isometric view.

In order to determine the mechanical parameters (tensile strength, yield strength, Young's modulus), the sample was tested on the TecQuipment SM1002 (Nottingham, UK) tensile testing stand and TeQuipment VDAS software (version 5.6.2.9196) (Figure 4).

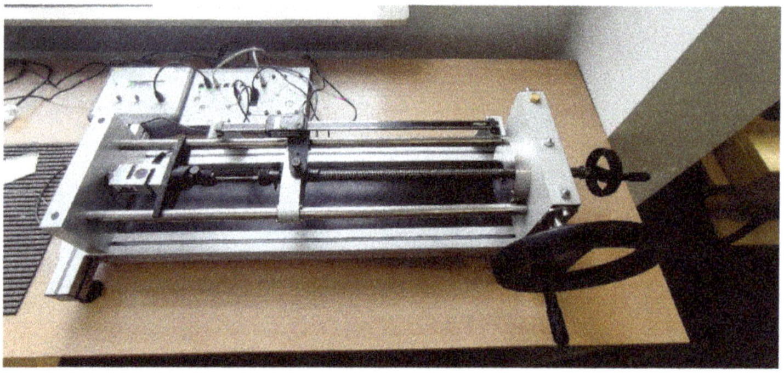

**Figure 4.** TecQuipment SM1002 Tensile Testing Station.

The tensile strength and the conventional yield strength of the sample were determined in the study. Based on the obtained results, the percentage elongation and the percentage reduction of the surface area were determined (Figure 5).

The determination of Young's modulus was carried out using an extensometer (Figure 6). For the collected data, a stress–strain graph was made and the Young's modulus was determined (Figure 7).

The collected data from both the tensile test and the determination of the Young's modulus allowed for determining the basic strength parameters of the material under consideration. The average tensile strength was determined to be approximately 714.58 MPa, the estimated yield strength was approximately 640 MPa, and the Young's modulus was estimated at 214.5 GPa. The percentage elongation of the sample in the tensile test was 5.7, while the reduction in the surface area was 58.7%.

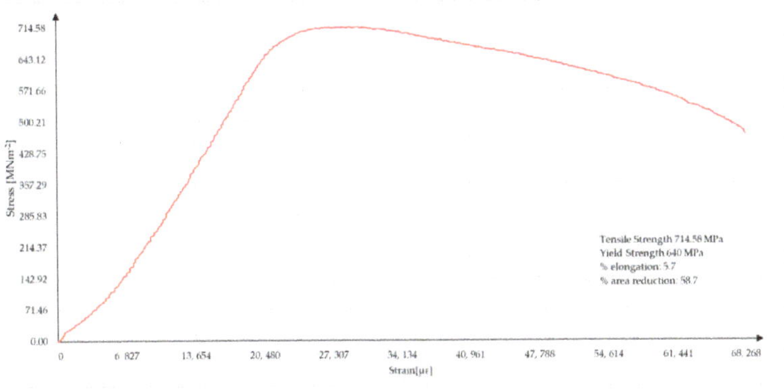


**Figure 5.** Stress–strain curve for the tested sample.

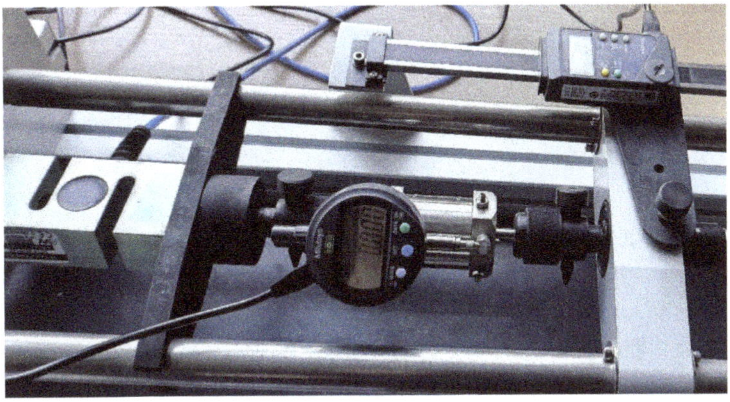

**Figure 6.** Sample with attached extensometer.

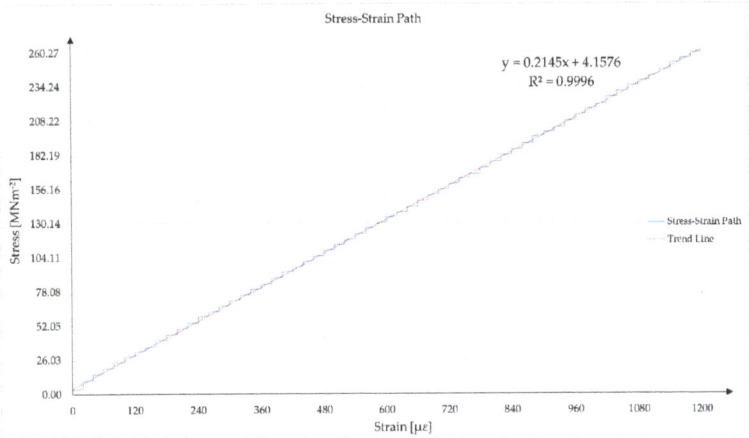


**Figure 7.** Stress–strain curve for Young's modulus determination.

## 2.2. Experimental Model of the Structure and Load

The test stand consists of two parts: a solid base with a load module made of rolled sections (I-beams and C-beams) and a ring made of square tubes on which the tested structure is based (Figure 8). Both parts are connected with clamps during the test.

**Figure 8.** Rod dome deflection test stand.

The structure is loaded using a hydraulic actuator. The hydraulic actuator is placed on a leveled pan. The pan is attached to a crane scale using a steel shackle. The currently applied load is read from the crane scale mounted in the keystone using a snap hook. The position of the hydraulic actuator is controlled in two directions using laser levels.

The tested structure is made of PO10 and PO12 drawn rods (Figure 9). The structure has rigid connections in the nodes, and the rods in the model made have been welded in the nodes. Nuts have been welded in the support nodes to secure the structure to the circumferential ring (the structure is bolted with one M12 bolt in each node, simulating an articulated support).

In order to accurately represent the existing structure for numerical analysis, the model was carefully measured after installation at the test site. The lengths of the bars in the nodes, the distances between the supports, and the height of each node (using a laser level as a reference plane for height) were measured.

During the test, displacements of the central node (keystone) and the support in node no. 1 were measured using vertical and leveled finger sensors. Five measurement series were carried out, and the results are the average values of all measurements.

The research was carried out as part of the National Science Centre, Poland, Miniatura 7 project (No. DEC-2023/07/X/ST8/00942) and constitutes preliminary research for further analyses.

(**a**)

(**b**)

**Figure 9.** Geometry of the rod dome (**a**) view, (**b**) top view.

### *2.3. Nonlinear Geometrical Analysis by Finite Element Method*

The choice of calculation method depends on the type of structure. The linear analysis (LA) is based on linear geometrical and physical relations between displacements, strains, and stresses. This analysis is reduced to the solution of a linear set of algebraic Equation (1):

$$\mathbf{K_L} \cdot \mathbf{q} = \mathbf{P} \tag{1}$$

where: $\mathbf{K_L}$—linear stiffness matrix of the structure, $\mathbf{q}$—vector of nodal displacements, $\mathbf{P}$—vector of nodal load.

Nowadays, the FEM software is usually equipped with a linear bifurcation analysis (LBA), which is to solve its own problem (2):

$$[\mathbf{K_L}+\mu\cdot\mathbf{K_G}]\cdot\mathbf{q} = \mathbf{0} \quad (2)$$

where $\mathbf{K_G}$—geometrical stiffness matrix, μ—load multiplier.

Unfortunately, shallow, single-layer domes are subject to large displacement gradients. These structures should be designed according to geometrically nonlinear analysis. The mathematical model of nonlinear FEM analysis corresponds to a set of nonlinear algebraic equations. We can write this system of equations in the incremental–iterative version as follows:

$$\mathbf{K_T}(\mathbf{q})\cdot\Delta\mathbf{q} = \Delta\mu\cdot\mathbf{P} + \mathbf{R} \quad (3)$$

where $\mathbf{K_T}(\mathbf{q})$—tangent stiffness matrix of the structure, $\mathbf{R} = \mathbf{P} - \mathbf{F}$—vector of residual forces, $\mathbf{F}$—vector of internal forces.

The tangent stiffness matrix $\mathbf{K_T}(\mathbf{q})$ is composed of linear stiffness matrix $\mathbf{K_L}$, stress stiffness matrix $\mathbf{K_{GN}}$, and nonlinear stiffness matrix displacements $\mathbf{K_{u1}}$ , $\mathbf{K_{u2}}$. In the equilibrium state $\mathbf{R} = \mathbf{0}$, while in the iterative process the standard $\mathbf{R}$ determines the distance from the equilibrium state. The iterative process converges if $\mathbf{R} \to \mathbf{0}$. In geometrically nonlinear problems we are dealing with a quadratic form of the eigenproblem:

$$[\mathbf{K_L}+\mu\cdot[\mathbf{K_{GN}} + \mathbf{K_{u1}} + \mathbf{K_{u2}}]]\cdot\mathbf{q} = \mathbf{0} \quad (4)$$

In this paper, the constant arc-length method developed by Riks [8,9] was applied to compute nonlinear equations using Abaqus software (Finite Element Method code).

## 3. Results and Discussion

A low-rise rod dome modelled as a frame structure is considered. The geometry of the dome was assumed based on the measured coordinates of the nodes of the constructed structure. The determined coordinates of the nodes are listed in Table 2. The structure is loaded at the central node (keystone). The material and strength data necessary for the numerical analysis were assumed in accordance with the performed experimental analyses (Section 2.1), i.e., Young's modulus E = 214.5 GPa, fy = 640 MPa.

**Table 2.** Node coordinates.

| No. of Node | X [m] | Y [m] | Z [m] | No. of Node | X [m] | Y [m] | Z [m] |
|---|---|---|---|---|---|---|---|
| 1 | 0.0 | −0.5280 | 0.1545 | 8 | −0.1480 | −0.2560 | 0.1155 |
| 2 | −0.5280 | −0.3050 | 0.1560 | 9 | −0.2960 | 0.0 | 0.1190 |
| 3 | −0.5280 | 0.3050 | 0.1590 | 10 | −0.1480 | 0.2560 | 0.1190 |
| 4 | 0.0 | 0.5280 | 0.1565 | 11 | 0.1480 | 0.2560 | 0.1150 |
| 5 | 0.5280 | 0.3050 | 0.1510 | 12 | 0.2960 | 0.0 | 0.1160 |
| 6 | 0.5280 | −0.3050 | 0.1505 | 13 | 0.0 | 0.0 | 0.1000 |
| 7 | 0.1480 | −0.2560 | 0.1150 | | | | |

### 3.1. Experimental Research

In the first stage, the actual structure was analyzed. The structure assembled as described in Section 2.2 was loaded with a manual hydraulic actuator. Table 3 presents the average force values read from the crane scale and the corresponding displacements. For each load, the displacement of the central node (keystone) and the vertical displacement of support node no. 1 were read.

**Table 3.** Experimental analysis results.

| Load [N] | Displacement [mm] [Node No.13] | Standard Deviation | Displacement [mm] Support [Node No. 1] | Standard Deviation |
|---|---|---|---|---|
| 0 | 0 | 0 | 0 | 0.0000 |
| 187.2 | 0.51 | 0.017 | −0.02 | 0.0050 |
| 276.8 | 0.83 | 0.035 | −0.01 | 0.0050 |
| 535 | 1.65 | 0.191 | 0.01 | 0.0082 |
| 717 | 2.27 | 0.200 | 0.02 | 0.0050 |
| 893 | 2.88 | 0.083 | 0.03 | 0.0129 |
| 1080 | 3.57 | 0.231 | 0.06 | 0.0050 |
| 1215 | 4.06 | 0.227 | 0.09 | 0.0096 |
| 1500 | 5.24 | 0.057 | 0.13 | 0.009 |
| 2007 | 7.94 | 0.067 | 0.23 | 0.005 |
| 2307 | 10.11 | 0.255 | 0.3 | 0.0129 |
| 2596 | 14.89 | 0.332 | 0.39 | 0.0096 |

### 3.2. *Numerical Analysis—'Standard'*

The second stage was the numerical analysis of the rod dome. A numerical model was created in the Abaqus program with analogous assumptions as in the case of the experimental structure (adopted geometry (Table 3), load scheme, material characteristics).

The dome was modelled as a frame structure (rigid joints were assumed). In the standard version, the structure bars are treated as ideal, without any deformations. It was assumed that the dome is based on hinge-resistant supports placed on an undeformable base.

The structure bars were divided 10 times to increase the accuracy of numerical calculations using the finite element method. The first step was linear buckling analysis. The structure was loaded in the keystone with a unit load. The critical load multiplier was 8.897. The first buckling form is presented in Figure 10. For the critical load multiplier value LBA < 10, the standard recommends performing nonlinear analysis.

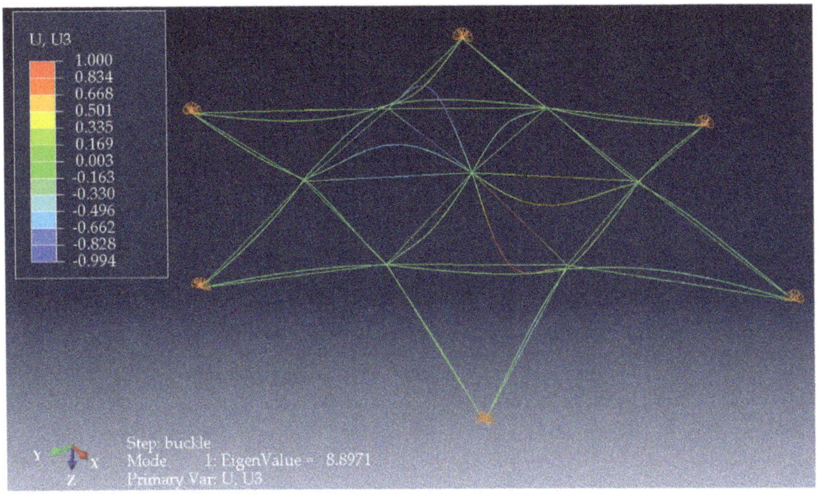


**Figure 10.** First form of buckling.

In the next step, a geometrically nonlinear analysis of the structure was carried out for loads analogous to those in the experimental study. The obtained results are presented in

Table 4. Figure 11 shows the equilibrium paths for the results obtained from experimental measurements and numerical calculations.

**Table 4.** Standard numerical analysis results.

| Load [N] | Displacement [mm] [Node No.13] |
|---|---|
| 0 | 0.000 |
| 187.2 | 0.212 |
| 276.8 | 0.318 |
| 535 | 0.638 |
| 717 | 0.874 |
| 893 | 1.111 |
| 1080 | 1.375 |
| 1215 | 1.573 |
| 1500 | 2.018 |
| 2007 | 2.922 |
| 2307 | 3.552 |
| 2596 | 4.277 |

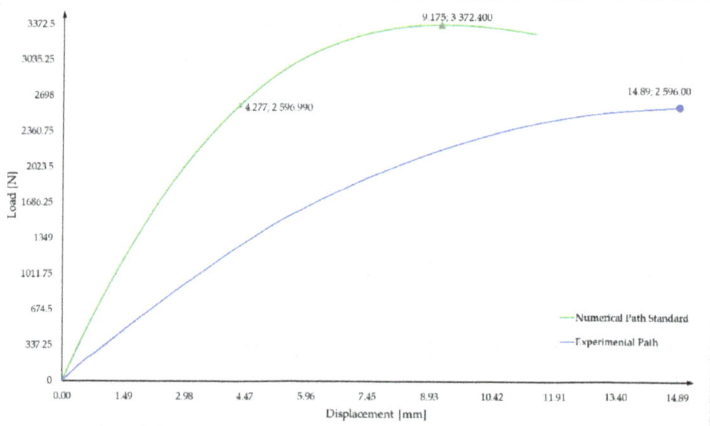


**Figure 11.** Equilibrium path for the keystone.

The determined values indicate significant differences in the results reaching more than 10 mm difference in displacements for the maximum loads. The authors decided to investigate what caused the situation. It was observed that the displacements of the support nodes were not equal to zero. The assumption regarding the non-deformable substrate is therefore incorrect. In the next calculation step, it was decided to take into account the determined elasticity of the supports based on the results contained in column 2 of Table 3.

*3.3. Numerical Analysis–Taking into Account Flexible Support*

After introducing the flexible support, the numerical model was changed. In each of the supports, the elasticity coefficients were taken into account with the following values: in the Z direction Kz = 1160 kN/m, in the X and Y directions Kx = Ky = 35,000 kN/m. The obtained results are presented in Figure 12 and Table 5.

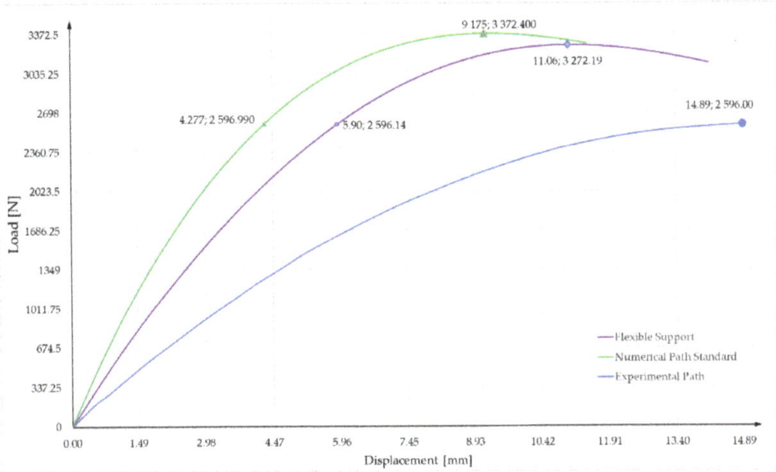


**Figure 12.** Equilibrium paths for the keystone.

**Table 5.** Numerical analysis results.

| Load [N] | Displacement [mm] [Node No.13] |
|---|---|
| 0 | 0.000 |
| 187.2 | 0.289 |
| 276.8 | 0.445 |
| 535 | 0.907 |
| 717 | 1.245 |
| 893 | 1.582 |
| 1080 | 1.953 |
| 1215 | 2.230 |
| 1500 | 2.843 |
| 2007 | 4.071 |
| 2307 | 4.916 |
| 2596 | 5.867 |

### *3.4. Numerical Analysis—Taking into Account Flexible Support and Imperfection*

In the next step, for the case with flexible supports, the influence of geometric imperfections was taken into account. Linear buckling analysis generates buckling modes of the structure and corresponding values of the critical force. Imperfections in Abaqus were introduced to the *Static Riks analysis, using the *imperfection command, as x% eigenvalues of the first buckling mode. For the realized structure, four imperfection values of 0.5, 1, 1.5, and 2% were considered. Figure 13 shows a comparison of values in relation to the experimental results. The best fitting curve is that corresponding to the construction with respect to 1% of the first eigenmode of LBA.

To evaluate the fit of the experimentally obtained graph to the numerically obtained graph, the mean square error (MSE) can be calculated (5):

$$MSE = \frac{1}{N}\sum_{i=1}^{N}\left(y_{exp}^{(i)} - y_{num}^{(i)}\right)^2 \quad (5)$$

where N—number of measurement points, $y_{exp}^{(i)}$—experimental value, $y_{num}^{(i)}$—numerical value.

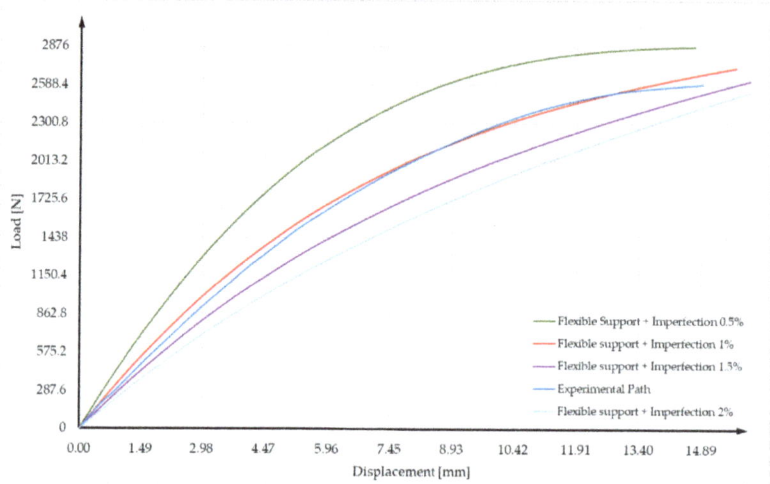


**Figure 13.** Equilibrium path for different values of imperfection allowance.

The mean square error between the experimental and numerical results taking into account flexible supports and imperfections of 0.5%, 1%, 1.5, and 2% were as follows: 13.76, 0.1, 2.07, and 5.56. The mean square error value indicates that the best fitting curve is the one with 1% imperfection value. Figure 14 shows a comparison of the equilibrium paths for all analyses performed.

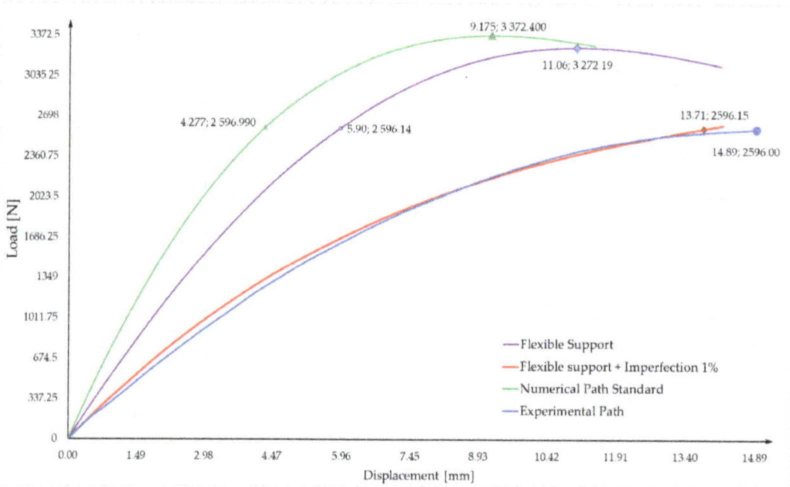


**Figure 14.** Equilibrium paths for the keystone.

### 3.5. Discussion

In this work, experimental results were verified with the results obtained from numerical analyses. The numerical model was developed based on the inventoried real structure.

The obtained results indicate significant differences in displacement values between the considered cases for analogous load values. The differences between the results from the standard analysis and the experimental ones reach over 10 mm of displacement for the maximum assumed load. Relative errors for the analyzed displacement results are presented in Table 6. The maximum determined error value determined for the load of

2596 N between the experimental and numerical results was 71.3%. It concerns the situation of modelling the reference structure, for which nominally rigid nodal connections and nominally pinned non-sliding supports were taken into account. As part of the experimental tests, it was observed that the supports were subject to displacements, and therefore in the next step the coefficient of elasticity of the supports was estimated. Taking into account the flexibility of the supports influenced the numerical results, reducing the maximum difference in displacement in the keystone to about 9 mm. However, the differences between the experimental and numerical results taking into account the elasticity of the supports are still noticeable. The relative errors for such a solution range from about 43% to about 60%.

**Table 6.** Relative errors in displacement.

| | Relative Error | | |
|---|---|---|---|
| **Load [N]** | **Experimental—Standard Analysis [%]** | **Experimental—Numerical Analysis with Flexible Supports [%]** | **Experimental—Numerical Analysis with Flexible Supports Taking into Account 1% of Imperfection [%]** |
| 187.2 | 58.4 | 43.3 | 4.5 |
| 276.8 | 61.7 | 46.4 | 11.8 |
| 535 | 61.3 | 45.0 | 10.5 |
| 717 | 61.5 | 45.2 | 10.5 |
| 893 | 61.4 | 45.1 | 8.1 |
| 1080 | 61.5 | 45.3 | 7.7 |
| 1215 | 61.3 | 45.1 | 7.7 |
| 1500 | 61.5 | 45.7 | 3.9 |
| 2007 | 63.2 | 48.7 | 1.5 |
| 2307 | 64.9 | 51.4 | −1.2 |
| 2596 | 71.3 | 60.6 | 7.9 |

After taking into account 1% of the eigenvalues of the first buckling mode, the results obtained described the experimental results well. The largest error in displacements between the experimental and numerical results taking into account imperfections was about 11.8% for 276.8N.

## 4. Conclusions

The inclusion of imperfections in the numerical model of a low-rise bar covering was crucial for accurately representing the actual behaviour of the structure under loading. In the context of low-rise bar covering, the inclusion of imperfections is particularly important. They can significantly affect the local and overall stability of the structure. For low-rise bar covering, initial imperfections can lead to significant changes in the behaviour of the structure. Even small curvatures with respect to the initial geometry can reduce the load-bearing capacity of the roof. The inclusion of imperfections in the numerical model often requires the inclusion of nonlinear effects in both the geometry and the material context. The cross-sections of low-rise bars are often subjected to large deformations, which makes their behaviour in the post-critical phase strongly nonlinear. The inclusion of initial imperfections can lead to significant deformations that would not be predicted in linear models. For this reason, in order to obtain realistic results, it is necessary to use more advanced models that take into account both nonlinear material properties and time-varying member characteristic curves. Including imperfections in the numerical model leads to more realistic results, especially in the case of nonlinear analyses. Thanks to such corrections, it is possible to obtain more accurate predictions of the behavior of the structure under the influence of loads, which is crucial for the design of safe structures. In the next stages of the analyses, it is planned to take into account material imperfections and modal analysis.

**Author Contributions:** Conceptualization, P.Z. and K.R.-K.; methodology, P.Z., K.R.-K. and P.G.K.; software, P.Z. and K.R.-K.; validation, P.Z., K.R.-K. and P.G.K.; formal analysis, P.Z. and K.R.-K.; investigation, P.Z. and K.R.-K.; resources, P.Z. and P.G.K.; data curation, P.Z. and K.R.-K.; writing—original draft preparation, P.Z. and K.R.-K.; writing—review and editing, P.Z., K.R.-K. and P.G.K.; visualization, P.Z. and K.R.-K.; supervision, P.G.K.; project administration, P.Z., K.R.-K. and P.G.K.; funding acquisition, P.G.K. All authors have read and agreed to the published version of the manuscript.

**Funding:** The tests were implemented thanks to the financial support of the Kielce University of Technology within the framework of the statutory work No. 02.0.20.00/1.02.001, SUBB. BKTK. 24.002.

**Data Availability Statement:** The original contributions presented in this study are included in the article. Further inquiries can be directed to the corresponding author.

**Acknowledgments:** This research was funded by National Science Centre, Poland 'Miniatura 7' project, grant number DEC-2023/07/X/ST8/00942.

**Conflicts of Interest:** The authors declare no conflicts of interest.

## References

1. Marcinowski, J. *Stability of Elastic Structures: Bar Structures, Arches, Shells*; Dolnośląskie Wydawnictwo Edukacyjne: Wrocław, Poland, 2017.
2. Waszczyszyn, Z.; Cichoń, C.; Radwańska, M. *Stability of Structures by Finite Element Methods*; Elsevier: Amsterdam, The Netherlands, 1994.
3. Kollár, L.P.; Tarján, G. *Mechanics of Civil Engineering Structures*; Woodhead Publishing: Sawston, UK, 2021.
4. Nordmark, A.; Eriksson, A. Structural Stability and Its Computational Modelling. In *Computational Methods in Engineering & the Sciences*; Springer: Berlin/Heidelberg, Germany, 2024.
5. Mironchenko, A. *Input-to-State Stability: Theory and Applications*; Springer: Berlin/Heidelberg, Germany, 2023.
6. Rust, W. *Non-Linear Finite Element Analysis in Structural Mechanics*; Springer: Berlin/Heidelberg, Germany, 2015.
7. Sinan, M. *Finite Element Method: Physics and Solution Methods*; Academic Press: Cambridge, MA, USA, 2022.
8. Riks, E. An incremental approach to the solution of snapping and buckling problems. *Int. J. Solids Struct.* **1979**, *15*, 529–551. [CrossRef]
9. Riks, E. The application of Newton's method to the problem of elastic stability. *J. Appl. Mech.* **1972**, *39*, 1060–1065. [CrossRef]
10. Ramalingam, R.; Jayachandran, S.A. Postbuckling behavior of flexibly connected single layer steel domes. *J. Constr. Steel Res.* **2015**, *114*, 136–145. [CrossRef]
11. Plaut, R.H. Snap-through of shallow reticulated domes under unilateral displacement control. *Int. J. Solids Struct.* **2018**, *24–34*, 148–149. [CrossRef]
12. Guan, Y.; Lawrence, N.V.; Daniel, H. Structural behavior of shallow geodesic lattice domes. *Int. J. Solids Struct.* **2018**, *155*, 225–239. [CrossRef]
13. Ligarò, S.S.; Valvo, P.S. Large displacement analysis of elastic pyramidal trusses. *Int. J. Solids Struct.* **2006**, *43*, 4867–4887. [CrossRef]
14. Rezaiee-Pajand, M.; Rajabzadeh-Safaei, N. Exact post-buckling analysis of planar and space trusses. *Eng. Struct.* **2020**, *223*, 111146. [CrossRef]
15. Xu, Y.; Han, Q.H.; Parke, G.A.R.; Liu, Y.M. Experimental Study and Numerical Simulation of the Progressive Collapse Resistance of Single-Layer Latticed Domes. *J. Struct. Eng.* **2017**, *143*, 04017121. [CrossRef]
16. Yan, S.; Zhao, X.; Rasmussen, K.J.R.; Zhang, H. Identification of critical members for progressive collapse analysis of single-layer latticed domes. *Eng. Struct.* **2019**, *188*, 111–120. [CrossRef]
17. Kala, Z. Computation of Equilibrium Paths in Nonlinear Finite Element Models. *MATEC Web Conf.* **2016**, *7*, 04026. [CrossRef]
18. Hrinda, G.A. Snap-Through Instability Patterns in Truss Structures. In Proceedings of the Conference: 51st AIAA/ASME/ASCE/AHS/ASC Structures, Structural Dynamics, and Materials Conference 18th AIAA/ASME/AHS Adaptive Structures Conference 12th, Orlando, FL, USA, 12–15 April 2010.
19. Silva, W.T.M.; Ribeiro, K.Q. Spatial asymmetric/symmetric buckling of Mises truss with out-of-plane lateral linear spring. *Int. J. Non-Linear Mech.* **2021**, *137*, 103810. [CrossRef]
20. Opatowicz, D.; Radoń, U.; Zabojszcza, P. Assessment of the Effect of Wind Load on the Load Capacity of a Single-Layer Bar Dome. *Buildings* **2020**, *10*, 179. [CrossRef]
21. Zabojszcza, P.; Radoń, U. Stability analysis of the single-layer dome in probabilistic description by the Monte Carlo method. *J. Theor. Appl. Mech.* **2020**, *58*, 425–436. [CrossRef]
22. Zabojszcza, P.; Radoń, U. Effect of Increased Density of Nodes in Geodesic Dome on its Critical Load Capacity. *IOP Conf. Ser. Mater. Sci. Eng.* **2019**, *471*, 052051. [CrossRef]
23. Radoń, U.; Zabojszcza, P.; Sokol, M. The Influence of Dome Geometry on the Results of Modal and Buckling Analysis. *Appl. Sci.* **2023**, *13*, 2729. [CrossRef]

24. Jankowski, J.; Kotełko, M.; Ungureanu, V. Numerical and Experimental Analysis of Buckling and Post-Buckling Behaviour of TWCFS Lipped Channel Section Members Subjected to Eccentric Compression. *Materials* **2024**, *17*, 2874. [CrossRef]
25. Czechowski, L.; Kotełko, M.; Jankowski, J.; Ungureanu, V.; Sanduly, A. Strength analysis of eccentrically loaded thin-walled steel lipped C-profile columns. *Arch. Civ. Eng.* **2023**, *69*, 301–316.
26. Di, C.; Zhang, J.; Wang, F.; Zhang, Y. Numerical and Experimental Buckling and Post-Buckling Analyses of Sphere-Segmented Toroidal Shell Subject to External Pressure. *Metals* **2023**, *13*, 64. [CrossRef]
27. Liang, K.; Sun, Q. An accurate and efficient implementation of initial geometrical imperfections in the predictor–corrector reduced-order modeling method. *Comput. Math. Appl.* **2020**, *79*, 3429–3446. [CrossRef]
28. Magisano, D.; Garcea, G. Increasing the buckling capacity with modal geometric "imperfections" designed by a reduced order model. *Thin-Walled Struct.* **2022**, *178*, 109529. [CrossRef]
29. Rzeszut, K. Post-Buckling Behaviour of Steel Structures with Different Types of Imperfections. *Appl. Sci.* **2022**, *12*, 9018. [CrossRef]
30. *EN 1990:2002*; Eurocode—Basis for Structural Design. European Union: Luxembourg, 2002.
31. *EN 1993-1-1*; Eurocode 3: Design of Steel Structures—Part 1–1: General Rules and Rules for Buildings. European Union: Luxembourg, 2005.
32. Radoń-Kobus, K.; Madej, M.; Kowalczyk, J.; Piotrowska, K. Properties of Diamond-like tungsten-doped carbon coatings lubricated with cutting fluid. *Coatings* **2024**, *14*, 342. [CrossRef]
33. Piotrowska, K.; Madej, M.; Kowalczyk, J.; Radoń-Kobus, K. Surface roughness effects on the properties of silicon-doped Diamond-like Carbon coatings. *Coatings* **2023**, *13*, 1629. [CrossRef]

metals

MDPI

*Article*

# Crystal Chemistry at Interfaces Between Liquid Al and Polar SiC{0001} Substrates

**Changming Fang * and Zhongyun Fan**

Brunel Centre for Advanced Solidification Technology (BCAST), Brunel University London, Uxbridge UB8 3PH, UK; zhongyun.fan@brunel.ac.uk
* Correspondence: changming.fang@brunel.ac.uk

**Abstract:** Silicon carbide (SiC) has been widely added into light metals, e.g., Al, to enhance their mechanical performance and corrosion resistance. SiC particle-reinforced metal matrix composites (SiC-MMCs) exhibit low weight/volume ratios, high strength/hardness, high corrosion resistance, and thermal stability. They have potential applications in aerospace, automobiles, and other specialized equipment. The macro-mechanical properties of Al/SiC composites depend on the local structures and chemical interactions at the Al/SiC interfaces at the atomic level. Moreover, the added SiC particles may act as potential nucleation sites during solidification. We investigate local atomic ordering and chemical interactions at the interfaces between liquid Al (Al(*l*) in short) and polar SiC substrates using ab initio molecular dynamics (AIMD) methods. The simulations reveal a rich variety of interfacial interactions. Charge transfer occurs from Al(*l*) to C-terminating atoms ($\Delta q$ = 0.3e/Al on average), while chemical bonding between interfacial Si and Al(*l*) atoms is more covalent with a minor charge transfer of $\Delta q$ = 0.04e/Al. The prenucleation at both interfaces is moderate with three to four recognizable layers. The information obtained here helps increase understanding of the interfacial interactions at Al/SiC at the atomic level and the related macro-mechanical properties, which is helpful in designing novel SiC-MMC materials with desirable properties and optimizing related manufacturing and machining processes.

**Keywords:** ceramic–metal interfaces; liquid-Al/SiC interfaces; prenucleation; interfacial interactions; ab initio molecular dynamics

## 1. Introduction

Silicon carbide (SiC) is a ceramic material. It exhibits advantageous properties, such as high hardness, good fatigue resistance, high thermal conductivity, low coefficient of thermal expansion, high chemical inertia, and resistance to corrosion [1,2]. Thus, SiC has many applications, including in automotive water pump seals, bearings, pump components, and extrusion dies [1–4]. Moreover, SiC has a mass density of 3.2 g/cm$^3$ [2–4], which is comparable with that of Al (2.7 g/cm$^3$) [5]. It has thus been utilized as an additive into light metals, such as Mg and Al, to manufacture (nano-sized) SiC particle-reinforced metal matrix composites (SiC-MMCs). SiC-MMCs display excellent mechanical performance, high corrosion resistance, and high volume/weight ratios [6–9]. They have potential applications in aerospace, automotives, etc. [8–10].

Experimental efforts have been focused on the preparation of SiC particle-enhanced composites, e.g., Al metal matrix composites (Al/SiC-MMCs) with different SiC contents [8–11], developments of preparation techniques, and properties measurements [6–10]. Recently, experimentalists worked hard at machining Al/SiC-MMCs and found various phenomena related to Al/SiC interfaces, such as matrix deformation and debonding between Al and SiC substrates [9–11]. These macroscopic properties of Al/SiC-MMCs originate from local atomic arrangements and interfacial chemical bonding. Knowledge about local crystal chemistry at the interfaces between liquid Al metal and SiC substrate surfaces is therefore vital to understand their macro-mechanical properties and applications and to understand the related manufacturing and machining processes.

Along the [0001] orientation, the structures of the hexagonal (4H-, or 6H-) SiC forms consist of SiC double-atom layers which can be regarded as the building blocks [1,12]. Rich SiC polytypes originate from the different stackings of SiC blocks. Each atom is tetrahedrally coordinated by the other species: three in the double-atom layer and one in the neighboring layer. Both Si and C atoms in the structures satisfy the $sp^3$ bonding. The unusual SiC structure indicates that cleaving the structure between the SiC double-atom layers is easier for creating SiC{0001} substrate surfaces. In this way, the produced SiC{0001} substrate surface is terminated either by C or by Si. In chemistry, the electronegativity is 1.90 in Pauling scale for silicon and 2.55 for carbon, respectively. Such a large electronegativity difference results in the created SiC{0001} surfaces having a polar nature, making them unstable at ambient conditions [13]. Fortunately, it is different in (liquid) metals since the free electrons of the metal atoms balance the polarity of the substrate surfaces [14,15]. Meanwhile, it is expected that the orientational $sp^3$ bonding of the outmost Si and/or C atoms with one dangling bond has impacts on the atomic ordering in the liquid adjacent to the substrates and the interfacial chemical bonding.

Above its nucleation temperature, there is atomic ordering in the liquid adjacent to a solid substrate [14,15]. This phenomenon is referred to as prenucleation [16,17]. Prenucleation supplies a precursor for following nucleation, which is crucial for solidification [18]. Such atomic ordering relates to interfacial bonding. The atomic arrangements of liquid Al atoms adjacent to the substrates and interfacial chemical bonding remain during casting [14,16,17]. Interfacial chemical bonding largely determines the mechanical properties of SiC-MMCs. Information about the local structure and chemical interaction at the liquid/solid interfaces thus helps us understand not only the nucleation potency of, e.g., the SiC substrates, but also the mechanical and chemical performances of the cast products [9,19] and the Al matrix deformation and interface debonding between the Al and SiC while machining SiC-MMCs.

Experiments have been conducted to understand the local structure at Al/SiC interfaces using various techniques, including high-resolution electron microscopy (HR-TEM) [20–22]. The HR-TEM images provided significant information about the atomic arrangements at the interfaces for the prepared samples. However, the obtained knowledge is mainly on the averaged atomic arrangements and is far from complete. In this aspect, theoretical approaches, especially parameter-free first-principle/ab initio methods, have advantages. This has been evidenced by previous studies on the assorted interfaces between liquid metal and solid substrates [14,15,23,24]. There have been reports on first-principles calculations for the interfaces between solid Al and SiC{0001} substrates with 3C- [25], 4H- [26,27], or 6H-stackings [28–31]. The calculations revealed stronger chemical interactions between the Al- and C-terminated substrates than between Al- and Si-terminated interfaces. Wang and Chen [32] performed atomistic modeling on misfit dislocations at the interfaces between SiC and several metals, including Al, using the potentials derived from ab initio calculations. Recently, ab initio molecular dynamics (AIMD) simulations were performed for the interfaces between liquid Mg and SiC{0001} substrates, and atomic ordering was observed in the liquid Mg near the SiC substrates [33]. Moreover, the simulations found atomic vacancies at the Mg layer adjacent to the substrates. To date, there is no report on the local structure and chemical interactions at the interfaces between liquid Al and SiC substrates (in short, Al(*l*)/SiC interfaces). Here, we investigate the atomic ordering/prenucleation and interfacial chemical bonding at the Al(*l*)/SiC{0001} interfaces using AIMD methods. Electronic band structure calculations are performed for equilibrated configurations. Bader charge analysis is employed to obtain the interface charge transfer between the liquid Al atoms and the substrate atoms based on the electron density distribution at the interfaces from the band structure calculations. This study reveals a rich variety of atomic ordering in the liquid Al atoms adjacent to the substrates and interfacial chemical interactions. There is a notable charge transfer from Al(*l*) to C-terminating atoms. Meanwhile, interfacial Si and Al($l$) atoms exhibit more covalent interactions. The information obtained here helps us understand the interfacial interactions of Al/SiC at the atomic level, which relates to the macro-mechanical properties of SiC-MMC materials and helps us design novel MMCs and related manufacturing and machining processes.

## 2. Methods

The ab initio approach employed in this study utilizes the periodic boundary conditions [34]. Thus, the usage of supercells is required. A hexagonal supercell was created with $a = 5a_0$, where $a_0$ is the length of the $a$-axis of the conventional hexagonal cell of 4H-SiC with consideration of the thermal expansion at the simulation temperature [33,35]. The $c$-axis is determined by the thickness of the SiC slab and the volume of the Al atoms with the density at the simulation temperature [5]. A supercell with $a$ = 15.53 Å and $c$ = 43.83 Å for the Al($l$)/SiC{0001} interfaces was obtained accordingly. This cell contains 400 Al($l$), and the substrate contains 100 Si and 100 C. The input interface model is shown schematically in Figure 1a. Such a sizable supercell helps attain statistically meaningful results and avoid the risks of artificial crystallization in the liquid metal.

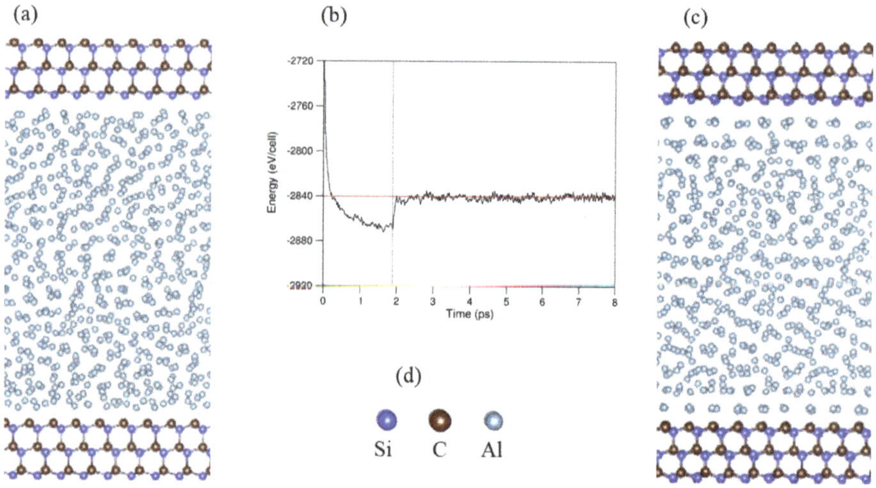


**Figure 1.** A schematic of the input structure (**a**), the relation between the total valence electron energy and simulation duration (**b**), and a snapshot of L-Al/SiC{0001} interfaces equilibrated at 1000 K (**c**). The meaning of the spheres (**d**) for (**a**,**c**) is as follows: the blue spheres represent Si, dark brown represents C, and silver represents Al. The liquid Al atoms adjacent to the substrates exhibit atomic ordering, and they are solid-like.

A plane-wave pseudo-potential path within the first-principle's code VASP was utilized [34,36]. This code employs the electronic density functional theory within the projector-augmented wave framework [37]. The generalized gradient approximation created by Perdew, Burke, and Ernzerhof (GGA-PBE) [38] was used for the exchange and correlation terms. The cut-off energies are $E_{CUT}/E_{AUG}$ = 520.0 eV/700.0 eV. These values are higher than the default values of the atoms ($E_{MAX}/E_{AUG}$ = 240.3 eV/291.1 eV for Al, 245.3 eV/322.1 eV for Si, and 400.0 eV/644.9 eV for C, respectively) for electronic structure calculations.

For the AIMD simulations, a cut-off energy of 320.0 eV was employed. This energy value is reasonable compared with the $E_{MIN}$ values of the pseudopotentials of the related atoms. The Γ-point in the Brillouin zone was used [39]. The latter is due to the lack of periodicity of the whole Al($l$)/SiC system. The AIMD simulations were carried out utilizing the NVT ensemble [34,36].

Liquid Al samples were produced first by equilibrating for about 3 ps (picoseconds) at 3000 K. Then, they were quenched to the designed temperature. The obtained liquid Al samples were inserted into the space between the SiC{0001} surfaces. A vacuum of about 2.0 Å was created between the surface of the liquid and the substrates (Figure 1a). AIMD simulations were performed via a two-step method. First, AIMD simulations were carried out with the substrate atoms pinned for over 1000 steps (1.5 fs (femtoseconds) per step). Then, the simulations were continued with full relaxation of the atoms for another 4000 steps. Analysis found no notable changes in atomic ordering at the interfaces. This two-step path prevents possible collective atomic movements which may occur when all atoms are relaxed from the start. The samples of the interfaces over 3.0 ps were used to ensure the results were statistically meaningful [34,40].

## 3. Results

Figure 1a shows the schematic structure of the input Al(*l*)/SiC interfaces, the model of which is described in the above section. There is a space between the solid substrates and the liquid Al at each side. The Al atoms display no long-range ordering. The ab initio calculations reveal a high total valence electron energy at $t = 0$ (Figure 1b).

When the simulation starts, the liquid Al atoms move around. The Al atoms at the borders move towards the fixed substrate atoms, gradually forming chemical bonding with the atoms at the substrate surfaces. Consequently, the energy decreases quickly in the first 0.3 ps and then decreases more smoothly until it reaches a constant value at about 1.6 ps (Figure 1b). Releasing the substrate atoms raises the energy dramatically. The energy reaches the equilibrium value within another 0.5 ps. Then, the energy varies around the equilibrium value. We equilibrated the system for another 5 ps. During the AIMD simulations, the substrate atoms vibrate around their equilibrium positions, remaining solid-like. An analysis revealed that from about 3.0 ps, the overall structure of the Al atoms remains unchanged. A snapshot of the equilibrated Al(*l*)/SiC{0001} system at 1000 K is presented schematically in Figure 1c.

Figure 1c shows obviously that the substrates' atoms and the (liquid) Al atoms are well-separated. The substrate atoms are well ordered and behave like they are in a solid phase at the simulation temperature. Meanwhile, the Al away from the interfaces displays no long-range ordering, and it is liquid-like. The Al atoms adjacent to the substrates exhibit localization, which decreases with the distance from the substrate surfaces. A closer look at Figure 1c shows the atomic density changing with an increasing distance from the surfaces. This phenomenon that atomic density varies with distance from the substrate in the direction perpendicular to the substrates is defined as layering. Layering can be described using the atomic density profile (ADF) [16,17,33,41]. An analysis for the equilibrated configurations over 3 ps was carried out, and the results are shown in Figure 2a. Variations in the peak heights for the layers around the interfaces are shown in Figure 2b.

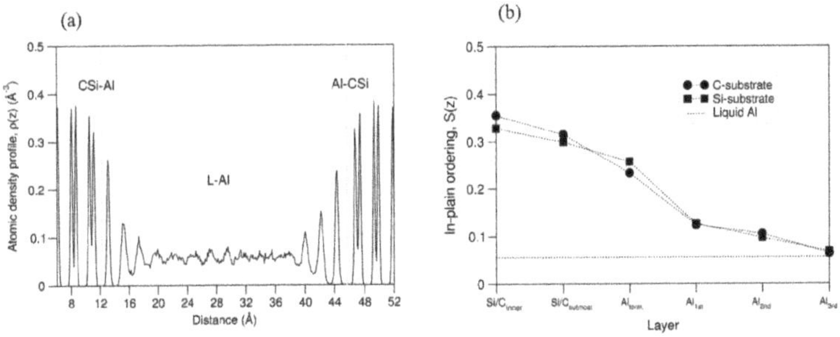


**Figure 2.** The atomic density profile, ρ(z) (**a**), and the variation in the peak heights of atomic layers (**b**). The straight dotted straight in (**b**) shows the average Al(*l*) density. The Al atoms adjacent to the SiC substrates form sharp layers.

Figure 2a shows sharp peaks for the Si and C atoms in the substrates, corresponding to their solid-like nature (Figure 1c). The termination Al atoms at both interfaces form peaks at 13.1 Å and 44.3 Å, respectively. These peaks are well separated from the neighboring layers. The ADF peak heights of the Al layers decrease with an increasing distance from the interfaces. The density of the third Al layer at both interfaces is just above the average Al($l$) value (Figure 2b). There are three recognizable Al peaks near the Si substrate. Near the C substrate, there are three to four apparent Al($l$) peaks, but the fourth peak is rather weak (Figure 2a). This indicates that layering phenomenon at the Al($l$)/SiC{0001} interfaces (three to four layers) is moderate. The analysis also showed that the peak distance between the outmost C-layer and the Al$(l)_{term}$ layer is 1.8 Å, which is shorter than that between the outmost Si-layer and the terminating Al layer (2.0 Å).

In the heterogeneous nucleation of light metals, solid grows layer by layer [42]. The atomic ordering in the liquid metals adjacent to the substrates acts as a template for the nearby liquid to nucleate. The atomic arrangements in the layers near the substrates for the equilibrated configurations over 3 ps are analyzed and shown in Figure 3.

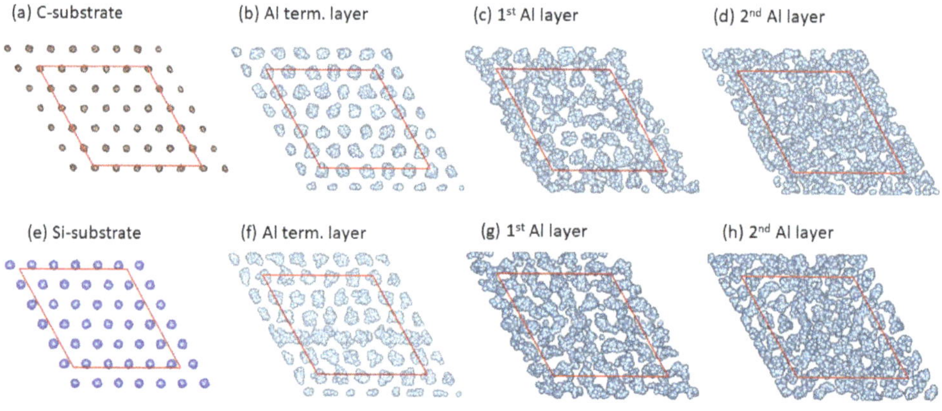


**Figure 3.** Averaged atomic arrangements of the outmost substrate layer and the Al layers at the C-terminated (**a–d**) and the Si-terminated (**e–h**) substrates over 3 ps. The red lines represent the in-plane axis for the supercells. The blue spheres represent Si, dark brown represents C, and silver represents Al. The terminating Al atoms (**b**,**f**) are highly ordered. The ordering of the terminating Al atoms at the C-terminating interface (**b**) is more significant than that at the Si-terminated interface (**f**).

To assess the ordering of the atoms in the layers near the substrates, an in-plane ordering coefficient [14,15,33,41] was employed for the configurations over 3 ps. The results are plotted in Figure 4.

From Figures 3 and 4, we obtained the following results:

(i) Both the surface C (Figure 3a) and Si (Figure 3e) atoms in the substrates are well ordered and solid-like. Their in-plane ordering coefficients are close to each other and to those in the substrates (Figure 4).

(ii) The terminating Al atoms at both substrates are highly ordered (Figure 3b,f), which corresponds well to the sharp ADP peaks and sufficient separation from the neighboring atoms (Figure 1b,c and Figure 2a). There are some subtle differences in atomic arrangements in the two terminating Al layers. The terminating Al atoms on the C substrate are separated from each other and show a localized nature, while some of the terminating Al atoms above the Si-terminated substrate show higher mobility and are connected to each other in the averaged arrangements (Figure 3f). Correspondingly, the in-plane ordering coefficient of the terminating Al atoms at the Si-terminated substrate is about 0.22, which is lower than that at the C-terminated substrate (0.31). The first Al layers at both substrates (Figure 3c,g) exhibit moderate atomic ordering, with the in-plane ordering coefficients being about 0.03.

There is little atomic ordering for the Al atoms at the second layer at both substrate surfaces, as shown in Figure 3d,h and Figure 4.

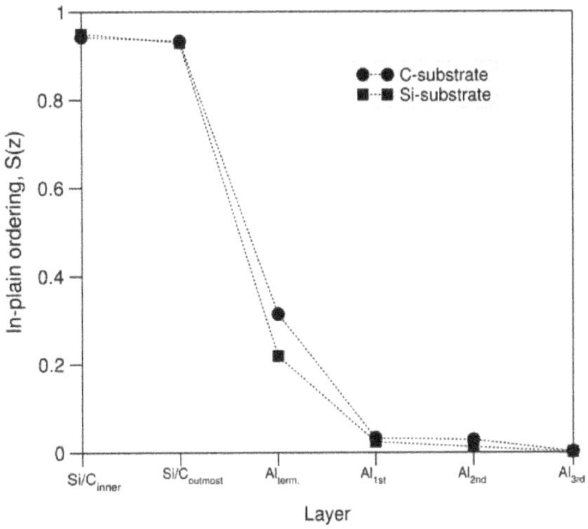


**Figure 4.** The in-plane ordering coefficients for the atomic layers nearby the L-Al/SiC{0001} interfaces. The $S(z)$ value for the terminating Al atoms at the C-terminated interface is higher than that at the Si-terminated interface.

We make statistics of atoms at each layer. The atomic ratio between those at the Al layer ($n$(Al)) and the substrate layer ($n$(C)), $n$(Al)/$n$(C), is 1.00 for the terminating Al layer, 1.05 for the first Al layer, and 1.13 for the second Al layer for the C-substrate. At the Si-terminated substrate, the corresponding ratios of $n$(Al)/$n$(Si) are 1.03, 1.06, and 1.13 for the terminating Al and the first and second Al layers, respectively. The atomic density of the second Al layer is close to the bulk values of the liquid Al misfit between $a$[SiC{0001}]/$a$[Al{111}] ~ 1.06 at the melting temperature of liquid Al, which corresponds to an area ratio of 1.13 [5].

The above results indicate stronger interfacial chemical bonding between C-Al than between Si-Al. The value of the atomic ratios at the interfaces is about 1, indicating that the majority of the outmost C/Si atoms have one Al neighbor, forming distorted tetrahedral coordination. Here, we analyzed the interfacial chemical interactions and chemical bonding in more detail. The atomic arrangements and local bonds at the C-substrate (Figure 5a) and the Si-substrate (Figure 5b) and the typical local coordination of the surface C (Figure 5c) and Si (Figure 5d) atoms are schematically presented, respectively.

Figure 5 shows that the terminating C or Si atoms have one Al atom as the nearest neighbor, forming a distorted tetragonal coordination which satisfies the $sp^3$ hybridization together with the three Si/C neighbors in the subsurface layer, according to Pauling's theory [43].

To obtain a deeper understanding of the local chemical interactions at the interfaces, we performed electronic band structure calculations for the Al($l$)/SiC system. The obtained electron density distributions are shown in Figure 6a. Based on the electron density distributions, we also analyzed the charges at the atomic sites at the interfaces using Bader's charge analysis model [44]. The obtained charges at the atomic sites at the interfaces are shown in Figure 6b. Moreover, we plotted the partial density of states for selected atoms at the interfaces in Figure 7.

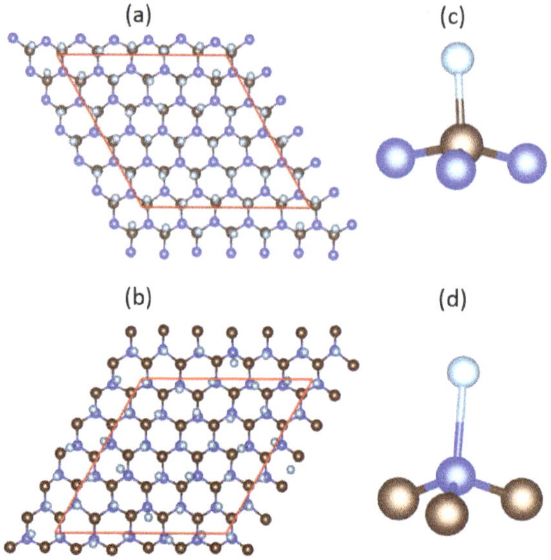


**Figure 5.** The atomic ordering of the terminating Al layer and the SiC surface layers with the C-termination substrate (**a**) and the Si-termination substrate (**b**) and the typical local coordination of the surface C (**c**) and Si (**d**) atoms. The red lines in (**a**,**b**) represent the in-plane axis. The blue spheres represent Si, dark brown represents C, and silver represents Al.

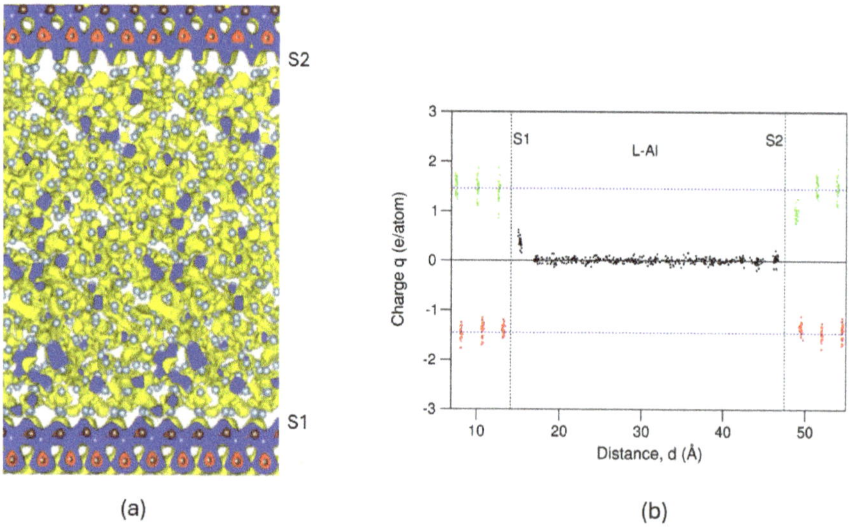


**Figure 6.** Electron density iso-surfaces with $\rho_0(r) = 0.035e/Å^3$ (**a**) and related Bader charges at the atomic sites (**b**). In 6a, the yellow clouds represent the iso-surfaces. The blue color represents regions with higher electron densities, and white represents regions with lower electron densities. The red regions represent the core regions of intensive electron densities. The blue spheres represent Si, dark brown represents C, and silver represents Al. In (**b**), the red dots represent charges at C, the green represents charges at Si, and black represents charges at Al. The charge transfer at the C-terminated interface (S1) is notably higher that at the Si-terminated interface (S2). All the snapshots and electron density iso-surfaces were produced via the program VASTA [45].

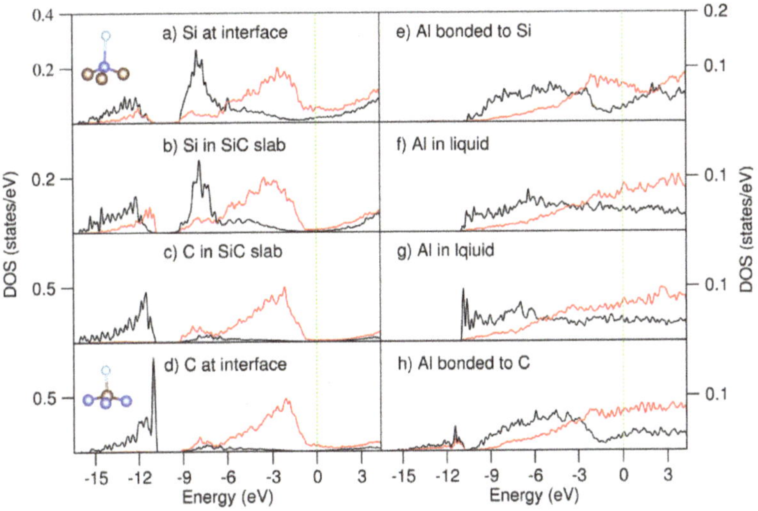


**Figure 7.** The partial density of state (pDOS) curves of selected atoms in the L-Al/SiC{0001} interfaces. The black curves represent the s-characters, and red represents the p-characters.

Figure 6a shows the localized electrons around the Si/C atoms in the substrates, while the Al atoms show its naked ions with (free) electrons around them. This corresponds to the strong Si-C covalent bonding and the free-electron nature of Al metals. Figure 6b shows the results of interatomic charge transfer from the Bader analysis. The Si atoms in the substrate are positively charged with a loss of about 1.5e/Si, whereas the C atoms obtained the same amount of electrons. The Al atoms away from the substrates are electronically neutral.

Figure 6b shows the occurrence of interfacial charge transfer (0.3e/Al on average) from the terminating Al atoms to the C-terminated substrate. Meanwhile, the terminating Al atoms lost little electrons (<0.1e/Al on average) to the outmost Si atoms. It is also noted that the outmost Si atoms lost fewer electrons than those in the substrate bulk. Those results are in good agreement with the electronegativity values, which are 2.5 for C, 1.8 for Si, and 1.5 for Al in Pauling's scale. The moderate charge transfer between Al and Si atoms corresponds to their more covalent interactions (the difference in electronegativity values is 0.3) compared with that between C and Al (the difference is 1.0).

Figure 7 shows the curves of the partial density of states for selected atoms around the interfaces. Around the Fermi level (zero eV), the electron densities of the Si (Figure 7b) and C (Figure 7c) in the center of the substrate slab are very low and more like pseudo-gaps. This corresponds to the semiconducting nature of SiC [4]. Meanwhile, there are nonignorable densities of states around the Fermi level for the outmost Si (Figure 7a) and C (Figure 7d). An analysis revealed that they originate from interfacial chemical interactions. Figure 7d shows a sharp peak at the top of the C 2 s states at −11.0 eV, which arises due to the C-Al interaction, as shown in Figure 7h, where there are some Al 3 s 3p states at the same energy range. Figure 7f,g show free-electron-like DOS curves for the Al atoms in the center of the Al slab. Even the partial density of states curves for the Al near the outmost Si (Figure 7a) shows weak densities in the low Si 3 s, 3p part between −15.5 eV and −11.0 eV. This indicates weak chemical bonding between the Al and Si atoms. The notable densities of states of the Si atom (Figure 7a) are the tails of the neighboring Al atoms.

## 4. Discussion

The AIMD simulations and electronic band structure calculations provided information about the local structures and interfacial interactions at the Al(*l*)/SiC{0001} interfaces. Each of the terminating C atoms is bonded to one Al, and thus, every C atom terminating the substrate has four (three Si and one Al) neighbors, fulfilling $sp^3$ hybridization. The interfacial interactions between Al and the terminating C atoms are strong with a charge transfer of about 0.3e/Al from Al to C (as summarized in d in Table 1). Therefore, energy costs for debonding at the Al(*l*)/SiC$\{0001\}_C$ interfaces will be high.

This study also showed anisotropy of chemical bonding at the interfaces: the bond between Al and the Si atoms terminating the substrate is relatively weaker and covalent compared to that between C and Al. There is a minor charge transfer from the terminating Al atoms to the Si atoms terminating the substrate (0.04e/Al on average; see e in Table 1). Moreover, the simulations also showed mobility of the terminating Al atoms, and there are some extra Al atoms in the terminating Al layer, with the $n$(Al)/$n$(Si) ratio being about 1.03. Such weaker Si-Al bonding means easier interfacial debonding at the Al/SiC$\{0001\}_{Si}$ interfaces during machining.

Prenucleation at a liquid metal/solid substrate interface is associated with the intrinsic capability (potency) of the substrate [46,47]. There are four factors affecting prenucleation at a solid metal/liquid substrate interface at a temperature above the nucleation point, including (i) the temperature, $T$ [17]; (ii) the lattice misfit between the substrate and the metal, $f$ [16,47]; (iii) the interfacial chemical interaction [17]; and (iv) the atomic roughness of the substrate, $R$ [24,46–48]. The interfacial interaction can be scaled by charge transfer at the interfaces [24,49].

The present study gives us an opportunity to address the impacts of these factors of the polar substrates on prenucleation in liquid light metals adjacent to the substrates. We compare the present results of the Al(*l*)/SiC interfaces with the interfaces of polar substrates: Mg(*l*)/SiC{0001} [33] and Al(*l*)/AlN [50] in Table 1. An idealized non-polar Al(*l*)/Al{111} interface [17] is included in Table 1 as a reference. All of these systems were simulated at the same temperature ($T$ = 1000 K).

The Al(*l*)/Al(s) interface is idealized and non-polar [17]. It does not contain a lattice misfit, atomic roughness, or an interfacial charge transfer (a in Table 1). The prenucleation at this interface is pronounced. There are six recognizable Al layers. The in-plane coefficient of the terminating Al layer is as high as 0.50.

The AlN{0001} substrates are polarized with either an Al-(g in Table 1) or N-terminated (f in Table 1) surface [50]. The lattice misfit is large: $f = -7.0\%$. The AIMD simulations provide strong anisotropy of the interfacial interaction and prenucleation at the interfaces. There are extra Al atoms at the terminating Al(*l*) layers at both Al-terminated (10%) and N-terminated (2%) interfaces, which correspond to an atomic roughness of 5.0% for the former and 1.0% for the latter (Table 1). Moreover, there are significant differences in interfacial interactions: there is a notable charge transfer from the terminating Al(*l*) atoms to the N in the substrate (0.52e/Al), which is opposite to the case at the Al-terminated substrate in which the terminating Al(l) atoms receive some electrons from their neighbors. The results indicate strong interfacial bonding at the Al(*l*)/AlN$\{0001\}_N$ interface and weaker bonding at the Al(*l*)/AlN$\{0001\}_{Al}$ interface. Consequently, layering at the former is strong with six recognizable layers, while it is moderate at the latter (four layers). The strong layering at the Al(*l*)/AlN$\{0001\}_N$ interface indicates a weak effect of the lattice misfit on layering, agreeing with the atomistic model [16]. The larger atomic roughness at the Al(*l*)/AlN$\{0001\}_{Al}$ interface causes it to have weaker layering. The in-plane ordering for the terminating Al layer at the Al(*l*)/AlN$\{0001\}_N$ interface is moderate (0.18) due to the large lattice misfit and atomic roughness [16]. More seriously is the in-plane-ordering at the Al(*l*)/AlN$\{0001\}_{Al}$ interface diminishing, which is caused by the combination of the large lattice misfit, atomic roughness [16,48,49], and the weak interlayer interactions [17].

Here, we compared the interfacial interactions and prenucleation of Al(*l*) and Mg{*l*} at the same SiC substrates and the Al(*l*)/SiC{0001} and Mg(*l*)/SiC{0001} [50] interfaces (Table 1). The large atomic size of Mg causes a positive lattice misfit (about 6.0%) to SiC{0001} (b and c in Table 1), whereas the lattice misfit between Al{111} and SiC{0001} is about -6.0% (d and e in Table 1). Table 1 shows that the charge transfers at the Mg(*l*)/SiC{0001} are notably higher than the corresponding ones at the Al(*1*)/SiC{0001} interfaces, indicating a more ionic nature at the former interfaces. This is due to the higher electronegativity value (1.61 in Pauling scale) of Al than that of Mg (1.31). Interestingly, there are notable amounts of atomic vacancies at the terminating Mg(*l*) layers, namely 5.0% for Mg(*l*)/SiC{0001}$_C$ (b in Table 1) and 15.0% for Mg(*l*)/SiC{0001}$_{Si}$ (c in Table 1) compared with that at the Al(*l*)/SiC{0001} interfaces (d and e in Table 1). This means there are stronger effects of the atomic volume at the Mg(*l*)/SiC{0001} interfaces than at Al(*l*)/SiC{0001}. Table 1 also shows that the recognizable layers in the liquid metals at the C-terminated substrates are slightly higher than those at the Si-terminated substrates. This indicates that stronger interfacial chemical interactions enhance layering. The numbers of recognizable layers in the liquid Mg (five to six) are higher than the corresponding one in the liquid Al (three to four). This partially comes from the larger Mg interlayer space, as shown in the pronounced layering at the Mg(*l*)/Zr interfaces (seven recognizable layers) in liquid Mg [51].

**Table 1.** A list of the lattice misfits between the substrates and the light metals, the substrates' surface features, and prenucleation (the number of ordered layers, $n_{Layers}$; the in-plane-ordering of the terminating metal layers, $S_M(z)$, for the interfaces between light metals and polar substrates, Al(*l*)/SiC{0001} (this work)) (d,e), Al(*l*)/AlN{0001} reprinted from Ref. [47] (f,g), and Mg(*l*)/SiC{0001} reprinted from Ref. [33] (b,c). The related properties for the idealized L-Al/s-Al{111} interface reprinted from Ref. [17] (a) were included as a reference. * The outmost Al atoms bonded to N are classified as the terminating atoms in the AlN substrates, while they were formerly defined as the 1st Al layer in reprinted from Ref. [50].

| Interface | *f* (%) | M Oc.(%) | *R*(%) | *q*(e/M) | $n_{Layers}$ | $S_M$(z) | Prenucleation |
|---|---|---|---|---|---|---|---|
| a: Al(*l*)/s-Al{111} [17] | +0.0 | 100 | 0.0 | 0.00 | 6 | 0.50 | Strong layering<br>Strong in-plane ordering |
| b: Mg(*l*)/SiC{0001}$_C$ [33] | +5.9% | 95 | 2.5 | +0.50 | 5 to 6 | 0.45 | Strong layering<br>Strong in-plane ordering |
| c: Mg(*l*)/SiC{0001}$_{Si}$ [33] | | 85 | 7.5 | +0.21 | 5 | 0.10 | Strong layering<br>Weak in-plane ordering |
| d: Al(*l*)/SiC{0001}$_C$ [This work] | −6.0% | 100 | 0.0 | +0.30 | 3 to 4 | 0.31 | Weak layering<br>Moderate in-plane ordering |
| e: Al(*l*)/SiC{0001}$_{Si}$ [This work] | | 103 | 1.5 | +0.04 | 3 | 0.22 | Weak layering<br>Moderate in-plane ordering |
| f: Al(*l*)/AlN{0001}$_N$ [50] | −7.0% | 102 | 1.0 | +0.52 | 6 | 0.18 | Strong layering<br>Moderate in-plane ordering |
| g: Al(*l*)/AlN{0001}$_{Al}$ [50] | | 110 | 5.0 | −0.31 * | 4 * | 0.02 * | Strong layering<br>Weak in-plane ordering |

Table 1 shows the relation between in-plane ordering and atomic roughness. The interfaces with smaller atomic roughness values for the C-terminated substrates have higher in-plane ordering coefficients with the Si-terminated substrates. This agrees with the previous conclusions from atomistic molecular dynamics modeling [48].

Table 1 also shows the anisotropy of chemical interactions at the interfaces between liquid metals (Mg, Al) and the polar substrates. The charge transfers from the metals to the more electronic negative atoms in the substrates (C and N) are notably larger than the values corresponding to the less electronegative atoms in the substrates (Si and Al). The obtained charge transfer between C and Al is 0.5e/Al (b), which is close to that between N and Mg at the interfaces (f). Therefore, these interactions are strong and more ionic. Meanwhile, the charge transfer from Al to C at the Al(*l*)/SiC{0001}$_C$ interface has a value of 0.3e/Al (d). This means that the bonding between Al and C at Al(*l*)/SiC{0001}$_C$ is more moderate compared with the other two, (b) and (f). The terminating Al atoms obtained

0.3e/Al on average from the substrate Al atoms (case g in Table 1). This is caused by the strong polarity of the AlN substrate where the outmost AlN slab with Al termination donates electrons to the terminating Al atoms, which means there is strong chemical bonding between the liquid Al and the substrate in (g). The charge transfer between Mg and the Si in the substrates in (c) is still notable (0.2e/Mg). Meanwhile, the charge transfer is insignificant between the Al and Si substrates (e) in Table 1. This corresponds to the weak interaction between them from the electronic structure calculations (Figure 7). Thus, comparatively, the chemical interactions between the liquid Al and SiC substrates are moderate, indicating easier deformation and debonding between the Al matrix and the SiC substrates during machining, as shown in the literature [9–11].

Information about the chemical bonding between the light metals and the polar substrates (Table 1) is helpful to understand the wetting at the M(*l*)/solid interface. Stronger interface interactions mean smaller wetting angles. Thus, the wetting angles at the Al(*l*)/SiC interface are expected to be larger than those at the Mg(*l*)/SiC and Al(*l*)/AlN systems. However, one must also be alert that polar surfaces such as the SiC and AlN substrates are unstable at ambient conditions [13]. Such polar substrates can be stabilized by defects [52] or the absorption of small molecules [53]. This would impact wettings when added into liquid metal [54,55]. This topic deserves further investigation.

In brief, the present ab initio study revealed that the interfacial interactions at the C-terminated substrates are more significant than those at the Si-terminated substrates at the Al(*l*)/SiC{0001} interfaces. Prenucleation at the Al(*l*)/SiC{0001} interfaces is overall moderate due to the large lattice misfit and moderate interfacial interactions.

## 5. Conclusions

The present AIMD simulations for the interfaces between liquid Al and the polar SiC{0001} substrates provide the following results:

(i) The substrates and liquid Al at the Al(*l*)/SiC{0 0 0 1} interfaces are well separated.
(ii) The Al layers adjacent to the SiC{0001} substrates are flat with the absence of or a moderate content of atomic vacancies.
(iii) Overall prenucleation at the Al(*l*)/SiC{0001} interfaces is moderate.
(iv) There is a moderate charge transfer from the Al atoms to the outmost C (0.3e/Al), while no significant charge transfer occurs from Al to Si.
(v) The moderate interface interactions between the Al and SiC substrates indicate possible interface debonding during machining.

This study provides insights into the interfacial interactions between Al and SiC particles in and macro-mechanical performances of Al/SiC-MMCs and the role of SiC particles in the nucleation of Al metals/alloys [9,56] and other MMCs [57,58]. Furthermore, it is useful to design novel manufacturing and machining processes for nano-sized SiC-MMCs of desirable properties in practice.

**Author Contributions:** Conceptualization, C.F.; methodology, C.F.; software, C.F.; validation, C.F.; formal analysis, C.F.; investigation, C.F.; resources, C.F.; data curation, C.F.; writing—original draft preparation, C.F.; writing—review and editing, C.F. and Z.F.; visualization, C.F.; supervision, Z.F.; project administration, C.F.; funding acquisition, Z.F. All authors have read and agreed to the published version of the manuscript.

**Funding:** The financial support from EPSRC (UK) under grant number EP/V011804/1 and EP/S036296/1 is gratefully acknowledged.

**Data Availability Statement:** The original contributions presented in this study are included in the article. Further inquiries can be directed to the corresponding author.

**Conflicts of Interest:** The authors declare no conflicts of interest.

## References


1. Kimoto, T.; Cooper, J.A. Chapter 2: Physical Properties of Silicon Carbides. In *Fundamentals of Silicon Carbides Technology: Growth, Characterization, Devices and Applications*; John Wiley & Sons (Pte. Ltd.): Singapore, 2014; pp. 10–37.
2. Dresch, A.B.; Venturini, J.; Arcaro, S.; Mentedo, O.P.K.; Bergmann, C.P. Ballistic ceramics and analysis of their mechanical properties for armour applications: A review. *Ceram. Intern.* **2021**, *47*, 8743–8761. [CrossRef]
3. Haynes, W.M. (Ed.) *CRC Handbook of Chemistry and Physics*, 97th ed.; CRC Press: Boca Raton, FL, USA, 2016; pp. 4–84.
4. Park, Y.-S. *SiC Materials and Devices*; Academic Press: Cambridge, MA, USA, 1988; pp. 20–60.
5. Arblaster, J.W. *Selected Values of the Crystallographic Properties of Elements*; ASM International: Materials Park, OH, USA, 2018.
6. Bekheet, N.E.; Gadelrab, R.M.; Salah, M.F.; Abd El-Azim, A.N. The effects of aging on the hardness and fatigue behavior of 2024 Al alloy/SiC composites. *Mater. Des.* **2002**, *23*, 153–159. [CrossRef]
7. Ahmed, A.; Neely, A.J.; Shankar, K.; Nolan, P.; Moricca, S.; Eddowes, T. Synthesis, tensile testing, and microstructural characterization of nanometric SiC particulate-reinforced Al 7075 matrix composites. *Metall. Mater. Trans. A* **2010**, *41*, 1582–1591. [CrossRef]
8. Atrian, A.; Majzoobi, G.H.; Enayati, M.H.; Bakhtiari, H. Mechanical and microstructural characterization of Al7075/SiC nanocomposites fabricated by dynamic compaction. *Int. J. Miner. Metall. Mater.* **2014**, *21*, 295–303. [CrossRef]
9. Wang, Y.L.; Monetta, T. Systematic study of preparation technology, microstructure characteristics and mechanical behavhors for SiC particles-reinforced metal materials composites. *J. Mater. Res. Technol.* **2023**, *25*, 7470–7497. [CrossRef]
10. Chen, Z.G.; Ding, F.; Zhang, Z.C.; Liao, Q.Y.; Qiao, Z.; Jin, Y.; Chen, M.J.; Wang, B. A review on machining SiCp/Al composite materials. *Micromachines* **2024**, *15*, 107. [CrossRef]
11. Zhao, G.L.; Mao, P.C.; Li, L.; Iqbal, A.; He, N. Micro-milling of 65 vol% SiCp/Al composites with a novel laser-assisted hybrid process. *Ceram. Intern.* **2020**, *46*, 26121–26128. [CrossRef]
12. Katkova, M.R.; Nosov, S.S.; Faddeev, M.A.; Chuprunov, E.V. On classification of silicon carbide polytypes. *Crystallogr. Rep.* **1999**, *44*, 795–798.
13. Tasker, P.W. The surface energies, surface tensions and surface structure of the alkali halide crystals. *Philos. Mag. A* **1979**, *39*, 119–136. [CrossRef]
14. Fang, C.M.; Fan, Z. Prenucleation at the liquid-Al/$\alpha$-$Al_2O_3$ and the liquid-Al/MgO interfaces. *Comput. Mater. Sci.* **2020**, *171*, 109258. [CrossRef]
15. Fang, C.M.; Fan, Z. Atomic ordering at the interfaces between liquid Al and MgO: An ab initio molecular dynamics study. *Philos. Mag. Lett.* **2020**, *100*, 235–244. [CrossRef]
16. Men, H.; Fan, Z. Prenucleation induced by crystalline substrates. *Metall. Mater. Trans. A* **2018**, *49*, 2766–2777. [CrossRef]
17. Fang, C.M.; Men, H.; Fan, Z. Effect of substrate chemistry on prenucleation. *Metall. Mater. Trans. A* **2018**, *49*, 6231–6242. [CrossRef]
18. Fan, Z.; Gao, G.; Jiang, B.; Que, Z.P. Impeding nucleation for more significant grain refinement. *Sci. Rep.* **2020**, *10*, 9448. [CrossRef] [PubMed]
19. Zhong, Z.; Jiang, X.S.; Sun, H.L.; Wu, Z.X.; Yang, L.; Matamoros-Veloza, A. Recent research on the optimization of interfacial structure and interfacial interaction mechanisms of metal matrix composites: A review. *Adv. Eng. Mater.* **2024**, 2401392. [CrossRef]
20. Lee, J.-C.; Ahn, J.-P.; Shim, J.-H.; Shi, Z.L.; Lee, H.-I. Control of the interface in SiC/Al composites. *Scr. Mater.* **1999**, *41*, 895–900. [CrossRef]
21. Liu, P.; Wang, A.-Q.; Xie, J.-P.; Hao, S.-M. Characterization and evaluation of interface in SiCp/2024 Al composite. *Trans. Nonferrous Met. Soc. China* **2015**, *25*, 1410–1418. [CrossRef]
22. Wang, Z.; Wang, A.Q.; Han, H.H.; Xie, J.P. HRTEM study of interfacial structure in SiCp/A390 composites. *Mater. Res. Express* **2020**, *7*, 046514. [CrossRef]
23. Wang, J.S.; Horsfield, A.P.; Schwingenschlögl, U.; Lee, P.D. Heterogeneous nucleation of solid Al from the melt by $TiB_2$ and $Al_3Ti$: An ab initio molecular dynamics study. *Phys. Rev. B* **2010**, *82*, 184203. [CrossRef]
24. Fang, C.M.; Fan, Z. Ab initio molecular dynamics investigation of prenucleation at liquid- Metal/Oxide Interfaces: An overview. *Metals* **2022**, *12*, 1618. [CrossRef]
25. Liu, B.B.; Yang, J.F. Mg on adhesion of Al(111)/3C-SiC(111) interfaces from first-principles study. *J. Alloys Compd.* **2019**, *791*, 530–539. [CrossRef]
26. Xu, X.Y.; Wang, H.Y.; Zha, M.; Wang, C.; Yang, Z.Z.; Jiang, Q.C. Effects of Ti, Si, Mg and Cu additions on interfacial properties and electronic structure of Al(111)/4H-SiC(0001) interface: A first-principles study. *Appl. Surf. Sci.* **2018**, *437*, 103–109. [CrossRef]
27. Li, S.; Arsenault, R.J.; Jena, P. Quantum chemical study of adhesion at the SiC/Al interface. *J. Appl. Phys.* **1988**, *64*, 6246–6253. [CrossRef]
28. Wu, Q.J.; Xie, J.P.; Wang, A.Q.; Ma, D.Q.; Wang, C.Q. First-principles calculations on the structure of 6H-SiC/Al interface. *Mater. Res. Express* **2019**, *6*, 065015. [CrossRef]
29. Wang, C.Q.; Chen, W.G.; Xie, J.P. Calculating study on properties of Al(111)/6H-SiC(0001) interfaces. *Metals* **2020**, *10*, 1197. [CrossRef]
30. Fathalian, M.; Postek, E.; Sadowski, T. Mechanical and electronic properties of Al(111)/6H-SiC interfaces: A DFT study. *Molecules* **2023**, *28*, 4345. [CrossRef]
31. Zhang, F.; Li, Q.; Glazoff, M.V.; Ott, R.T. First-principles study of interfaces in Al/SiC metal-matrix composite system. *Comput. Mater. Sci.* **2023**, *229*, 112444. [CrossRef]

32. Wang, D.; Chen, N.X. Atomistic study of misfit dislocation in metal/SiC(111) interfaces. *J. Phys. Condens. Matter* **2010**, *22*, 135009. [CrossRef]
33. Fang, C.M.; Fan, Z. Prenucleation at L-Mg/SiC{0001} interfaces from ab initio molecular dynamics simulations. *Metall. Mater. Trans. A* **2023**, *51*, 788–797. [CrossRef]
34. Kresse, G.; Hafner, J. Ab initio molecular dynamics for liquid metals. *Phys. Rev. B* **1993**, *47*, 558–561. [CrossRef]
35. Talwar, D.N.; Sherbondy, J.C. Thermal expansion coefficient of 3C-SiC. *Appl. Phys. Lett.* **1995**, *67*, 3301–3303. [CrossRef]
36. Kresse, G.; Furthmüller, J. Efficiency of *ab-initio* total energy calculations for metals and semiconductors using a plane-wave basis set. *Comput. Mater. Sci.* **1996**, *6*, 15–50. [CrossRef]
37. Blőchl, P.E. Projector augmented-wave method. *Phys. Rev. B* **1994**, *50*, 17953–17978. [CrossRef]
38. Perdew, J.P.; Burke, K.; Ernzerhof, M. Generalized gradient approximation made simple. *Phys. Rev. Lett.* **1996**, *77*, 3865–3868. [CrossRef]
39. Monkhorst, H.J.; Pack, J.D. Special points for Brillouin-zone integrations. *Phys. Rev. B* **1976**, *13*, 5188–5192. [CrossRef]
40. Brostow, W.; Hagg Lodbland, H.E. *Materials: Introduction and Applications*; John Wiley & Sons: Hoboken, NJ, USA, 2017.
41. Hashibon, A.; Adler, J.; Finnis, M.W.; Kaplan, W.D. Atomistic study of structural correlations at a liquid-solid interface. *Comput. Mater. Sci.* **2002**, *24*, 443–452. [CrossRef]
42. Fan, Z. An epitaxial model for heterogeneous nucleation on potent substrate. *Metall. Mater. Trans. A* **2013**, *44*, 1409–1418. [CrossRef]
43. Pauling, L. The nature of the chemical bond. Application of results from the quantum mechanics and from a theory of paramagnetic susceptibility to the structure of molecules. *J. Am. Chem. Soc.* **1931**, *53*, 1367–1400. [CrossRef]
44. Bader, R.F.W.A. A quantum-theory of molecular-structure and its applications. *Chem. Rev.* **1991**, *91*, 893–928. [CrossRef]
45. Momma, K.; Izumi, F. VESTA 3 for three-dimensional visualization of Crystals, volumetric and morphology data. *J. Appl. Cryst.* **2011**, *44*, 1272–1276. [CrossRef]
46. Fan, Z. Heterogeneous Nucleation, Grain Initiation and Grain Refinement of Mg-Alloys. In Proceedings of the 11th International Conference on Magnesium Alloys and Their Applications, Sitges, Spain, 24–27 July 2018; Fan, Z., Mendis, C., Eds.; Beaumont Estate: Old Windsor, UK, 2018; p. 7.
47. Men, H.; Fang, C.M.; Fan, Z. Prenucleation at the Liquid/Substrate Interface: An Overview. *Metals* **2022**, *12*, 1704. [CrossRef]
48. Jiang, B.; Men, H.; Fan, Z. Atomic ordering in the liquid adjacent to an atomically rough solid substrate. *Comput. Mater. Sci.* **2018**, *153*, 73–81. [CrossRef]
49. Fang, C.M.; Fan, Z. Atomic ordering at the Liquid-Al/$MgAl_2O_4${111} Interfaces: Ab initio molecular dynamics simulations. *Metall. Mater. Trans. A* **2020**, *51*, 6318–6326. [CrossRef]
50. Fang, C.M.; Fan, Z. Atomic ordering at the interfaces between liquid aluminum and dipolar AlN{0001} substrates. *Metall. Mater. Trans. A* **2022**, *53*, 2040–2047. [CrossRef]
51. Fang, C.M.; Fan, Z. A Comparative Study of Prenucleation on Zr and MgO Substrates by ab initio MD simulations. In Proceedings of the 11th International Conference on Magnesium Alloys and Their Applications, Sitges, Spain, 24–27 July 2018; Fan, Z., Mendis, C., Eds.; Beaumont Estate: Old Windsor, UK, 2018; pp. 41–50.
52. Yu, L.; Zunger, A. A polarity-induced defect mechanism for conductivity and magnetism at polar-nonpolar oxide interfaces. *Nat. Commun.* **2014**, *5*, 5118. [CrossRef] [PubMed]
53. Macko, J.; Podrojková, N.; Oriňaková, R.; Oriňak, A. New insights into hydrophobicity at nanostructured surfaces: Experiments and computational models. *Nanomater. Nanotechnol.* **2022**, *12*, 18479804211062316. [CrossRef]
54. Xi, L.; Kaban, I.; Nowak, R.; Korpała, B.; Bruzda, G.; Sobczak, N.; Mattern, N.; Eckert, J. High-temperature wetting and interfacial interaction between liquid Al and $TiB_2$ ceramic. *J. Mater. Sci.* **2015**, *50*, 2682–2690. [CrossRef]
55. Xiao, H.Y.; Yu, Z.L.; Liang, J.C.; Ding, L.; Zhu, J.S.; Wang, Y.F.; Chen, S.G. Wetting behavior-induced interfacial transmission of energy and signal: Materials, mechanics, and applications. *Adv. Mater.* **2024**, *36*, 2407856. [CrossRef]
56. Lai, L.; Niu, B.; Bi, Y.; Li, Y.; Yang, Z. Advancements in SiC-reinforced metal matrix composites for high-performance electronic packaging: A review of thermo-mechanical properties and future trends. *Micromachines* **2023**, *14*, 1491. [CrossRef]
57. Gupta, S.; Sharma, A.K. Microstructure and Microhardness of Mg/SiC Metal Matrix Composites Developed by Microwave Sintering. *J. Inst. Eng. India Ser. C* **2022**, *103*, 63–68. [CrossRef]
58. Singh, A.; Singh, J.; Sinha, M.K. Ferrous-metal matrix composites: A review on status, scope and challenges. *Int. J. Interact. Des. Manuf.* **2023**, *17*, 2807–2829. [CrossRef]

metals

MDPI

*Article*

# Assessing Fatigue Life Cycles of Material X10CrMoVNb9-1 through a Combination of Experimental and Finite Element Analysis

**Mohammad Ridzwan Bin Abd Rahim [1,2,*], Siegfried Schmauder [1], Yupiter H. P. Manurung [3], Peter Binkele [1], Ján Dusza [4], Tamás Csanádi [4,5], Meor Iqram Meor Ahmad [6], Muhd Faiz Mat [3] and Kiarash Jamali Dogahe [1,7]**

1 Institute for Material Testing, Materials Science and Strength of Materials (IMWF), University of Stuttgart, 70569 Stuttgart, Germany; siegfried.schmauder@imwf.uni-stuttgart.de (S.S.); peter.binkele@imwf.uni-stuttgart.de (P.B.); kiarash.dogahe@imwf.uni-stuttgart.de (K.J.D.)
2 Department of Occupational Safety and Health (DOSH) Sarawak, Kuching 93100, Malaysia
3 Smart Manufacturing Research Institute (SMRI) and School of Mechanical Engineering, Universiti Teknologi MARA (UiTM), Shah Alam 40450, Malaysia; yupiter.manurung@uitm.edu.my (Y.H.P.M.); muhdfaizmat@uitm.edu.my (M.F.M.)
4 Institute of Materials Research of SAS, Watsonova 47, 040 01 Košice, Slovakia; jdusza@saske.sk (J.D.); tcsanadi@saske.sk (T.C.)
5 Donát Bánki Faculty of Mechanical and Safety Engineering, Óbuda University, Népszínhás utca 8, 1081 Budapest, Hungary
6 Department of Mechanical and Manufacturing Engineering, Faculty of Engineering and Built Environment, Universiti Kebangsaan Malaysia (UKM), Bandar Baru Bangi 43600, Malaysia; meoriqram@ukm.edu.my
7 Graduate School of Advanced Manufacturing Engineering (GSaME), University of Stuttgart, Nobelstraße 12, 70569 Stuttgart, Germany
* Correspondence: ridzwan.rahim@imwf.uni-stuttgart.de

**Abstract:** This paper uses a two-scale material modeling approach to investigate fatigue crack initiation and propagation of the material X10CrMoVNb9-1 (P91) under cyclic loading at room temperature. The Voronoi tessellation method was implemented to generate an artificial microstructure model at the microstructure level, and then, the finite element (FE) method was applied to identify different stress distributions. The stress distributions for multiple artificial microstructures was analyzed by using the physically based Tanaka–Mura model to estimate the number of cycles for crack initiation. Considering the prediction of macro-scale and long-term crack formation, the Paris law was utilized in this research. Experimental work on fatigue life with this material was performed, and good agreement was found with the results obtained in FE modeling. The number of cycles for fatigue crack propagation attains up to a maximum of 40% of the final fatigue lifetime with a typical value of 15% in many cases. This physically based two-scale technique significantly advances fatigue research, particularly in power plants, and paves the way for rapid and low-cost virtual material analysis and fatigue resistance analysis in the context of environmental fatigue applications.

**Keywords:** X10CrMoVNb9-1 (P91); two-scale material modeling (TSMM); Tanaka–Mura model (TMM); Paris law; fatigue life cycles

## 1. Introduction

Power plants have been generating massive quantities of electricity and refining raw materials into final products. The typical life span of a power plant is 402,960 h [1,2]. Low cycle fatigue stresses are imposed on machinery that operates at high temperatures and with some vibration due to its surroundings. Increasing a power plant's operating temperature within the allowed range is a principal means of improving energy efficiency, but it requires remarkable structural integrity [3]. Monitoring structural integrity and detecting local failures are critical for ensuring that respective components remain fit for service. Understanding a material's fatigue life under cyclic load is essential because structures can collapse under fatigue loading at loads lower than the material's yield stress.

Numerous researchers [4,5] have examined the X10CrMoVNb9-1 P91 steel's high-temperature crack development characteristics under constant load. Saad et al. investigated P91 materials to develop a constitutive viscoplasticity model capable of reproducing the mechanical behavior of power plant materials under thermomechanical fatigue conditions [6]. They discovered that viscoplasticity might be employed in predicting P91 steel behavior at temperatures ranging from 400 °C to 600 °C. Jinbiao et al. investigated the prediction of high-temperature fatigue crack onset in P91 steels [7]. They observed that high-angle grain boundaries were reinforced by 74% when they compared the fine-grain heat-affected zone alongside the coarse-grain heat-affected zone.

The accumulation of damage under cyclic loading is the key issue of fatigue, making it a nuanced phenomenon [8]. The fatigue life of engineering constructions is an intricate process with different phases. To simplify this procedure, engineering structure design often considers three superordinate phases: fracture initiation, fatigue crack propagation, and unstable crack growth. These phases are frequently evaluated separately. Figure 1 shows a graphic representation of the stage of the fatigue life cycle. Zhan investigated the fatigue behavior of TC4-TC11 titanium alloys through laser-melting deposition [9], while Huang et al. adopted a damage mechanics approach to examine the effects of overload on notched specimen fatigue, combining residual stress and plastic damage [10].

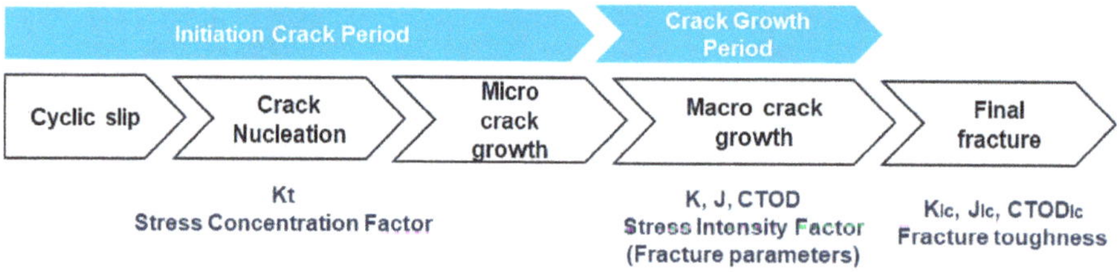


**Figure 1.** Life cycles of fatigue.

Several researchers [2,11–13] have employed the Tanaka–Mura model (TMM) as a modeling tool to determine the initiation of cracks. The concept was first presented by A. Brückner-Foit and X. Huang [14], who developed a computer model for the evolution of fracture due to fatigue in martensitic steels (F82H) using an updated version of the TMM. Brückner-Foit's research focuses more on quantifying cycles than does Kramberger et al.'s [13] investigation into the segmental crack nucleation process. Mlikota and Schmauder et al. [15] used the improved TMM to establish multiscale modeling and simulation of the metal fatigue method. The researchers estimated the number of cycles needed for the ultimate fracture using a dual-scale and a multiscale framework to assess crack initiation and propagation. They successfully evaluated the simulation of crack initiation with experiments on various steel varieties, including AISI 1141, high-strength steel S960, and martensitic steel.

Santus et al. investigated the development of small metal cracks [16]. They discovered that crack development has three stages. The first stage is the microstructurally short crack (MSC) stage, which raises questions about the applicability of continuum mechanics when the crack dimension is equivalent to or less than the grain size. The second stage is the physically short crack (PSC) stage, in which the crack barely grows due to reduced closure and other influences. The final stage is the long crack (LC) stage, in which the crack continues to develop following the Paris law until it ultimately fractures.

Following Tanaka and Mura's argument [17,18], a dislocation model that incorporates a double pile-up positioned on a single slip band effectively captures the inception phase of the crack. The interaction between forward and reverse loading results in fatigue, and over time, materials exhibit flaws, such as dislocations. As the cycles increase, new crystals show evidence of slip bands, and the existing bands widen, with some eventually evolving into short cracks. The slip band or slip lines initially appear in a few well-aligned crystals (grains) within the substructure as narrow and sharp lines. In most metals, the slip systems are numerous, and the active ones are oriented close to the planes with the maximum shear stress. When a force is applied in only one direction (uniaxial loading), the crack planes will always be angled at an estimated 45°.

The Paris law frequently illustrates the propagation of long cracks (as shown in Equation (1)). It develops a relationship with the stress intensity factor (SIF) $K$, which occurs at the crack's tip during the steady crack growth duration and the fatigue crack growth rate, which is also known as FCGR or $da/dN$ [19,20].

$$da/dN = C(\Delta K)^m \tag{1}$$

The constants obtained via experimentation are $C$ and $m$. The equation formally describes an experiment on the formation of fatigue cracks. An essential detail to recognize is that physical laws guide the development of fatigue cracks. The relationship between the crack growth rate ($da/dN$) and the stress intensity factor range ($\Delta K$) curve is where the Paris law is applied, as shown in Figure 2. The mechanism that causes the fracture to propagate is pushed by the crack driving force. This specific force is related to the value of the critical SIF $\Delta K_{Ic}$.

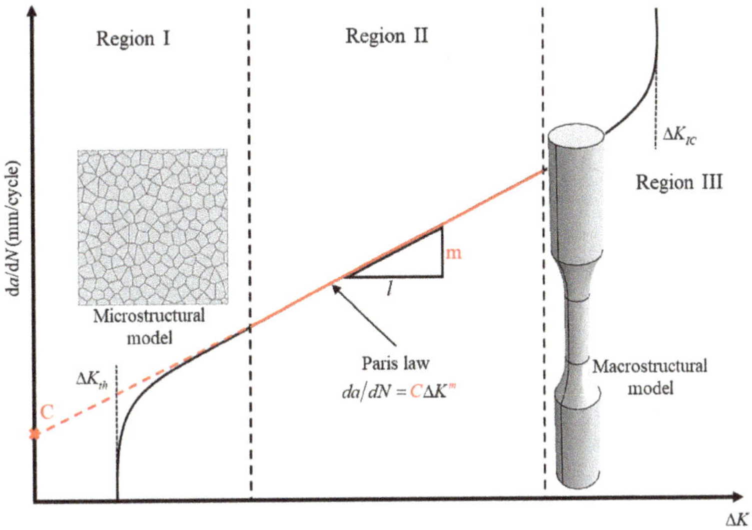


**Figure 2.** $da/dN$ vs. $\Delta K$ plot describing the three regions and influencing factors associated with crack growth rate.

In the microstructural study field, few investigations focus explicitly on the material X10CrMoVNb9-1 (P91) that analyze the initiation of short cracks in fatigue. The integration of the Voronoi tessellation (VT) technique with the TMM at the microstructure level, particularly for the P91 material, provides a novel approach to understanding the material's fatigue behavior, particularly regarding crack initiation and propagation in elastic–plastic modeling [7]. The steps of crack initiation and propagation are different, which is why the entire fatigue lifetime needs to be understood and the evaluation of the crack initiation and growing processes needs to be compared through experiments. The physical TMM [11] determines how many cycles a single grain needs to start to microcrack. This analytical model is evaluated numerically and is based on the physical principles of dislocation movement. Three artificial microstructure models were created for this investigation using the VT technique at room temperature. Simulation findings will be reviewed and discussed for conclusions and recommendations.

## 2. Methodology

### *2.1. P91 Material*

The P91 material is a corrosion-resistant steel, whose principal alloying components are approximately 0.1% carbon, 9.0% chromium, and 1.0% molybdenum. This structure includes micro-alloying components such as vanadium and niobium in addition to manganese and silicon. These substances produce MX thermally resistant carbides and carbonitride, particularly V (C,N) and Nb (C,N). At high temperatures, the high strength results from the interaction between these precipitates and dislocations in the matrix. P91 is typically utilized in its hardened state. M23C6 carbides stabilize the structure by locating themselves within grains and sub-grains [21–23]. The high chromium concentration of P91 is partly responsible for its excellent oxidation resistance. Through solid solution hardening, manganese and silicon contribute to an increase in mechanical strength. The chemical composition of the specifications and the investigated heat are detailed in Table 1.

**Table 1.** Mechanical characteristics of P91 material at ambient temperature.

| Mechanical Properties (Data from Ref. [24]) | 24 °C |
|---|---|
| Shear modulus, G (MPa) | 78,986 |
| Poisson's ratio, υ | 0.38 |
| Young's modulus, E (MPa) | 218,000 |
| Yield stress (MPa) | 550 |
| Ultimate tensile stress (MPa) | 690 |

Martensitic steel in its 9Cr-1Mo (P91) form is the substance used in this study, according to ASME Section II and the experimental results. Table 1 displays the material's mechanical properties [24], and Table 2 displays its chemical composition [4].

Electron backscatter diffraction (EBSD) (Figure 3b) was performed to observe the crystallographic orientation and shape of the grains of the microstructure of P91. Three representative microstructures were used to obtain better fatigue crack initiation results, as in Figure 3c.

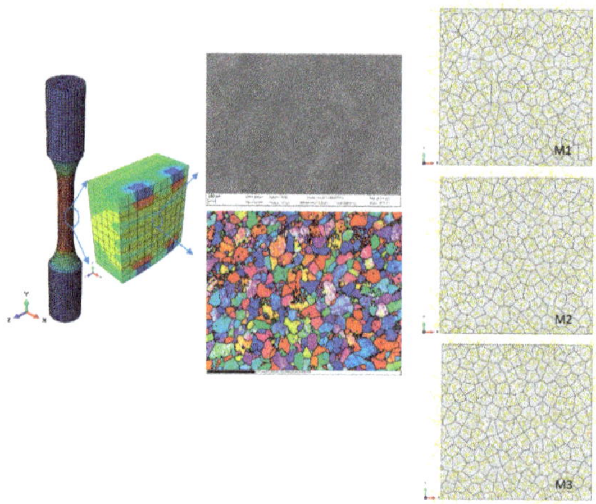


(a) (b) (c)

**Figure 3.** Two-scale models illustrating macrostructure as a global model and 3D submodel (**a**), SEM and EBSD image (**b**), and three artificial microstructures (submodel) derived from EBSD image (**c**).

**Table 2.** Chemical composition of P91 steel in wt%.

| | Chemical Composition in Mass (%) (Data from Ref. [4]) | | | | | | | | | | | | |
|---|---|---|---|---|---|---|---|---|---|---|---|---|---|
| | C | Si | Mn | P | S | Cr | Mo | Ti | V | Nb | Ni | N | Al |
| Minimum | 0.09 | 0.26 | 0.37 | 0.15 | 0.0 | 0.88 | 0.91 | 0.01 | 0.20 | 0.08 | 0.08 | 0.44 | 0.01 |

### 2.2. Experimental Procedure

The experimental approach was used to determine the number of fatigue cycles that X10CrMoVNb9-1 can withstand at room temperature. The specimens for the fatigue test were made of P91 steel (yield stress = 550 MPa; tensile strength = 690 MPa; elongation = 46%). Table 2 displays the chemical composition of the material. This steel is frequently used to create different structural components for power plants, such as superheaters and steam pipelines that can operate up to 600 °C. The samples were taken from a tube with a 57.0 mm diameter and 15.6 mm wall thickness.

The fatigue test was performed at ambient temperature and used constant strain and stress amplitudes. An Instron 8502, from INSTRON (Norwood, MA, USA), testing machine with additional parts for low-cycle testing at room temperature was used for these tests, Figure 4. The stress amplitude-controlled testing used four unique total stress amplitude levels, namely, 300, 425, 475, and 558 MPa. The parameters related to every fatigue test are listed in detail in Table 1. The loading frequency was held constant at 0.5 Hz across all experiments. Instantaneous loads were imposed on the specimens, and the consequent fatigue cycles were measured. The loading ratio (R) was set as −1 and the specimen dimensions as in Figure 5.

The procedure follows ASTM E606-04 standards for strain-controlled fatigue testing [25].

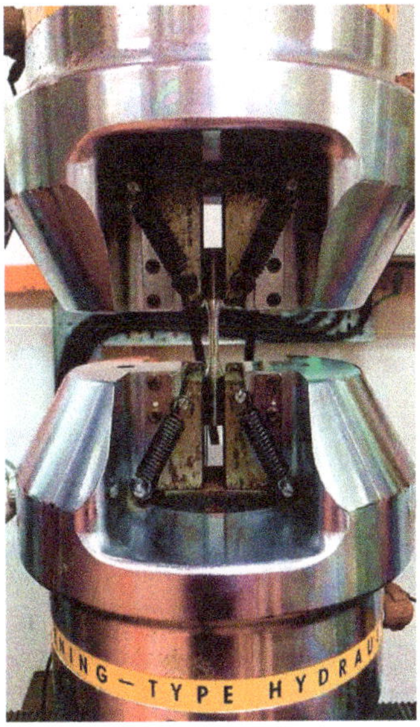


**Figure 4.** Specimen during the fatigue test.

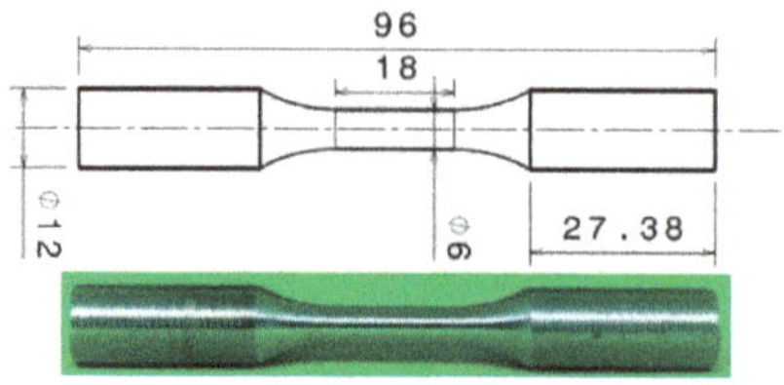


**Figure 5.** (**a**): Sketch and dimension (unit in millimeters); (**b**): actual experimental fatigue specimen.

### 2.3. Modeling Fatigue via Elastic–Plastic Analysis

TMM helps to determine how many cycles are needed for cracks to start and grow in individual grains. The crack-initiation stress-number (S-N) of cycles curves estimate how many cycles a crack will last before it breaks. The material being investigated in this study, 9Cr-1Mo (P91), has a body-centered cubic crystal structure [26] and is commonly utilized when high temperatures are involved. The fatigue life curve was derived by assessing the number of cycles required for the initiation of short cracks across varying microstructures. An enhancement over previous studies is the incorporation of plasticity, better aligning the model's characteristics with the real material properties. This adjustment is particularly relevant because it accounts for crack initiation, marking the onset of irreversible material deformation.

The deformation of the elastic–plastic model can be divided into an elastic part and an inelastic (plastic) part. This statement is written as Equation (2):

$$\mathbf{F} = \mathbf{F}^{el} \bullet \mathbf{F}^{pl} \tag{2}$$

where $\mathbf{F}$ is the total deformation gradient; $\mathbf{F}^{el}$ is the fully recoverable part of the deformation at the point under consideration ($\left[\mathbf{F}^{el}\right]^{-1}$ is the deformation that would occur if, after the deformation $\mathbf{F}$, inelastic responses were somehow prevented, but at the same time, the stress at the point was reduced to zero); $\mathbf{F}^{pl}$ is defined by $\mathbf{F}^{pl} = \left[\mathbf{F}^{el}\right]^{-1} \bullet \mathbf{F}$ [27]. We consider the slip mechanisms in the microscale as a constitutive model for crystal plasticity. Consequently, the critical resolved shear stress (CRSS) is incorporated into the TMM equation.

### 2.4. Fatigue Crack Initiation Model

In 1981, Tanaka and Mura developed a dislocation model [18,19] that included a dual pile-up formation on a single slip plane to explain how cracks nucleate. The number of fatigue crack initiations needs to be determined before calculating the total number of cycles for fracture due to fatigue. Crack nucleation inside a single grain can be predicted using the TMM, as indicated in Equation (3):

$$N_g = \frac{8GW_c}{\pi(1-v)d(\Delta\overline{\tau} - 2\text{CRSS})^2} \tag{3}$$

where $N_g$ is the number of cycles for fatigue crack initiation, $G$ is the shear modulus, $W_c$ is the specific fracture energy per unit area, $v$ is the Poisson's ratio, $d$ is the length of the slip band, $\Delta\overline{\tau}$ is the average shear stress range on the slip band, and CRSS stands for critical resolved shear stress, which is the critical value of shear stress in the glide direction, whereby the dislocation movement can occur. Grain boundary dislocation pile-up is prevented if the resolved shear stress is smaller than the CRSS.

### 2.5. Fatigue Crack Propagation

The growth rate of prolonged fatigue cracks (stated as $da/dN$) at particular stress conditions for a wide range of technical alloys is generally defined by the Paris law [28], which is a widely recognized theory based on linear elastic fracture mechanics (LEFM):

$$\frac{da}{dN} = C(\Delta K)^m \tag{4}$$

where $da/dN$ is the crack growth rate, $\Delta K = K_{\max} - K_{\min}$ is the SIF range at the crack tip during stable crack growth, and C and m are (the Paris) constants of the material that can be inferred via fitting to experimental data or computed mathematically, respectively [29].

Figure 11 shows how the $a - N$ curve can be utilized to determine the number of cycles needed for an LC to propagate, $N_{prop}$, spanning from the point at which crack initiation is complete, $a_0$, to the eventual failure point and $a_f$ of a specimen or component. This curve can be obtained by integrating a power-law equation, such as [30]:

$$N(a) = \frac{1}{C}\int_{a_0}^{a_f} \frac{da}{(\Delta K)^m} \tag{5}$$

In terms of the final failure, the $a - N$ curve and fracture toughness, $\Delta K$, have an impact such that $N_{prop} = f(Na, K_{Ic})$. The Paris law and its amendments [31] should be noted as the sole set of criteria that can be used to determine the durability of structural components with early cracks.

A numerically based S-N (Wöhler) curve can be created by integrating the results from the TMM and the Paris law (Equations (1) and (3)), and it can be used to predict when the final failure can be expected to happen for various loading levels.

### 2.6. Physical Model

A physical model represents a real model of an object or system. It is designed to represent, simulate, or predict the behavior of the actual object or system in the real world. This model is applied when conducting experiments or gathering data would be impractical, costly, or impossible. Numerical simulations of crack initiation were carried out using ABAQUS [32], which included a plug-in for the TMM and VT, to specifically address micro-crack nucleation. The ABAQUS scripting interface is used by the Python code that invokes the graphical user interface [33]. The execution, in this case, consists of two parts: Python programming, which manages and directs the simulation procedure, and ABAQUS, which handles tasks such as pre-processing, solving the actual finite element (FE) problems, and post-processing (Figure 6).

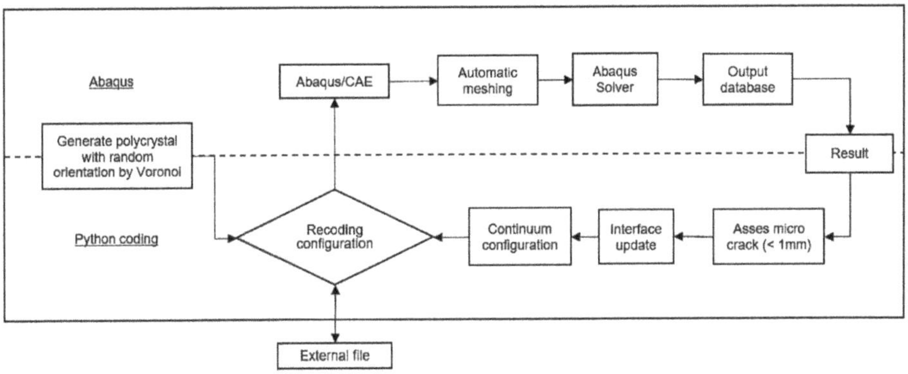


**Figure 6.** Algorithm for evaluating the number of initiation cycles during the crack initiation phase.

Figure 6 depicts the algorithm for evaluating fatigue crack initiation. The initial stage involves the development of a polycrystalline microstructure. This process requires four main parameters in VT, namely, seeds, perturbation, grain size, and boundary. Seeds are starting points for the initiation of grain structure, perturbation is a slight displacement to these seeds to ensure randomness and mimic natural microstructure variations, grain size is the average size of the grains, and grain boundaries refer to the outline of the simulated microstructure and the interface between grains. The micro-model is derived based on VT to simulate the assembly of polycrystalline grains. Each grain's displacement field in the micro-model, which collectively represents an infinite variety of orientations, is used to calculate the characteristics of the boundary interface. An auxiliary file is created with microcrack geometry, slip band, and slide segment properties for every grain. This model serves as the ground zero for the next phase of the crack initiation process.

The model is meshed using the Mesh module in ABAQUS, and the solver provides a database of displacements, strains, and stresses as outputs. The next microcrack in the micro-model can be predicted using a custom-developed routine (Python algorithm) using the shear stress magnitude on each segment. Grain boundaries—representing the grains within the microstructure—are said to have failed if a microcrack forms on an already present slip plane segment.

On the basis of the TMM in Equation (3), the average shear stress on a slip plane, $\Delta\overline{\tau}$, must exceed twice the $CRSS$ $(\Delta\overline{\tau} - 2CRSS)$ value. This will ensure the activation of the necessary slip systems to identify the most vulnerable segments. The equation's denominator $(\Delta\overline{\tau} - 2CRSS)$ provides an indirect justification for this condition. The TMM predicts a slip plane segment fracture that satisfies the criterion and requires the fewest cycles for nucleation. After a new crack that starts at a particular place is added, the microstructural model is remeshed, and the evaluation under further cyclic loading is repeated until no more segments can be cracked.

The microscale model shown in Figure 3c is used to analyze the initiation of fatigue cracks [13]. This model shows that microcrack nucleation models can reliably forecast the beginnings of fatigue cracks. The VT technique creates an extensive microstructure before this model can be used. On the basis of the experimentally acquired EBSD image in Figure 3b, the microstructure of the submodels (M1, M2, and M3) was produced randomly, with an average grain size of 20 μm. X. Yang et al. [34] and Li et al. [35] similarly found 20 μm grain sizes for 9Cr-1Mo at different temperatures.

Representative volume elements (RVEs), which are produced using the VT approach, are a key component of the simulation model. A random orientation is assigned to each cell in the Voronoi structure. Each cell represents a prior martensite grain, which cause the random angles of the cells; thus, the crystal structure orientations in the plane are defined [13]. The Voronoi method correctly represents the average grain size of a material's microstructure. The micro-models of this work typically comprise 104,501 distinct elements and an average of 199 grains. The 3D shell models possess a non-zero thickness and differ from their 2D equivalents in containing non-constant through-thickness stresses. This 3D shell submodel represents a layer of grains in the XY plane with a thickness equal to the average grain size (20 μm) (Figure 3c).

### 2.7. Evaluation of Short Crack Initiation Wöhler (S-N) Curve Utilizing Numerical Methods

The microstructural model estimates brief fatigue crack nucleation and propagation. Generating the micro-model for fatigue initiation analysis is crucial. This model is carefully placed at the area of maximal stress concentration where crack initiation is expected. In this work, 0.3 mm × 0.3 mm micro-models with an average grain size that mimics the initiation process were chosen. This scale was selected to balance computational feasibility with the model size that accurately represents physical crack initiation processes. These dimensions were enough to view the crack nucleation. Figure 3c shows the carefully designed artificial microstructure model based on the EBSD image. This scientific approach gives a granular view of material fatigue behavior, improving our understanding of crack initiation before propagation.

Although the micro-model is generated in 2D in Figure 3c, implementing the M3D4R elements makes the model a three-dimensional deformable shell mesh. This model uses membrane components and simplified integration techniques to reduce computational complexity. One common use of these elements is to model thin stiffening components within a continuum. These components usually act as reinforcing layers in various materials and structure systems.

The micro-models employed in this study exhibit orthotropic elastic–plasticity material properties, which allow an enriched understanding of material behavior, conforming to the requirements in Table 1. The incorporation of plastic properties within the microstructure model marks an enhancement over previous research [2]. This condition represents that the microstructure has direction-dependent properties toward the elastic region and undergoes plastic deformation. Equations (6) and (7) are used to obtain the elastic constants for cubic symmetry, which are then applied to the micro-models and are shown in Table 3. Notably, C44, as expressed in Equation (8), corresponds to the material's shear modulus. The CRSS and the crack initiation energy ($W_c$) are the two material parameters included in the TMM, which was used in this study. According to earlier research [35], the $W_c$ value can be derived from the $J$-$R$ curve as derived from experiments or from calculations of a CT specimen. The CRSS, set at 134.5 MPa for this study, was determined by experimental data, the specifics of which will be covered in a subsequent paper.

**Table 3.** Elastic constants for cubic crystal structures applied in the micro-model to characterize the material behavior.

| Elastic Properties | Equation of Elastic Constant for Cubic Crystal | | Value of Elastic Constant at 24 °C, MPa |
|---|---|---|---|
| $C_{11} = C_{22} = C_{33}$ | $C_{11} = \frac{E(1-v)}{(1-v-v^2)}$ | (6) | 408,092 |
| $C_{12} = C_{13} = C_{23}$ | $C_{12} = \frac{Ev}{(1-v-2v^2)}$ | (7) | 250,121 |
| $C_{44} = C_{55} = C_{66}$ | $C_{44} = G = \frac{Ev}{2(1+v)}$ | (8) | 78,986 |

### 2.8. Geometry of the Macro-Models

Developing a thorough macro-scale model known as the global model for the specimen is required to create a numerical representation for analyzing crack propagation. This model has two purposes: first, to convey the stress amplitude and as a boundary condition to the microstructure model, and second, to aid in the precise computation of the SIF for crack propagation analysis. The critical SIF would be achieved by adjusting the seam crack length $a$ on the model using the extended FE method (XFEM).

As demonstrated in Figure 7, the model utilized in this investigation is axisymmetric. Axisymmetric modeling is a type of FE analysis that can be performed within Abaqus under the presumption that the model's geometry, material properties, and loading conditions are all symmetric about some central axis. This method is also called 2.5D and makes computing much more efficient by letting a rotational symmetric three-dimensional object be shown and studied as a cross-section in two dimensions. A study on mesh convergence guided the determination of mesh density.

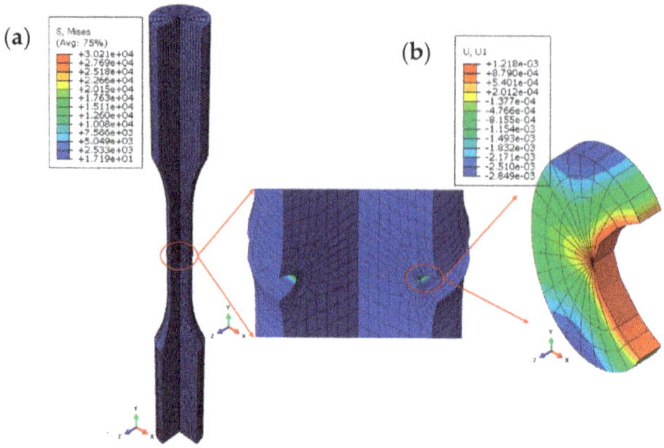


**Figure 7.** The 3D FE model, incorporating symmetrical boundary conditions, utilized to compute the SIF, $K$ for varying crack lengths: (**a**) in unit MPa and (**b**) in unit mm.

## 3. Result and Discussion

On the basis of the EBSD image in Figure 3b, three separate micro-models mirroring a particular microstructure were chosen for the numerical investigation of fatigue crack initiation lifetime. These models were subjected to seven stress amplitudes at room temperature: 300, 350, 400, 425, 475, 524, and 558 MPa. Figure 4 depicts the experimental results obtained by the School of Mechanical Engineering at UiTM Shah Alam, Malaysia, serving as a premise for comparison with the simulated outcomes. The fatigue test with stress amplitudes of 350, 400, and 524 MPa was not conducted experimentally due to the limitation of specimens and to observe the pattern of the number of cycles from TMM. The grains subjected to the maximum stress levels, typically the weakest grains expected to initiate the first MSCs, were identified by using an Abaqus plug-in enhanced with the TMM. Figure 9b shows the microcrack model in which the shear stresses change between grains and within each grain due to the orientation of each grain and the load applied. The TMM Equation (3) indicates that this result means the first microcrack starts along the grain slip line with the shortest expected lifetime.

The average shear stress range ($\Delta\overline{\tau}$) on the slip line, obtained from the FE analysis, is a crucial input in this equation. The local stress field changes once the crack is initiated, making it less tight. The number of cycles required to initiate a new crack is subsequently recalculated for each grain and slip line in the microstructure. Similar to the case where the microstructure is unaffected, the subsequent MSC that needs the fewest cycles to start is found. Then, a crack appears in the RVE. The model was improved to support the sequential nucleation of cracks at the segmental level, which aligns with the improvement described in [12]. In each simulation step, only one slip line segment of a specific grain develops a crack. If a segment from one grain breaks in one stage, then a segment from another grain may probably crack in the next step at the same cycle number or at a higher one. For this change, the TMM equation had to be changed by changing the grain-breaking cycles ($N_g$) to segment-breaking cycles ($N_s$) and the slip line length ($d$) with the slip line segment length ($d_s$) [36,37]. The model was remeshed after creating each segmental crack, and the process was then repeated.

For this investigation, each grain slip line was divided into four equal-sized segments. On the basis of the TMM equation, each MSC or segmental crack that forms in the micro-model has its length and number of life cycles. The crack growth rate is calculated by dividing these two output numbers. Figure 8b shows this crack growth rate when plotted against the number of cracked slip line segments. In all the studied cases, the crack growth rate oscillated and went down after a certain number of cycles. Figure 8b, which depicts the 300 MPa stress amplitude instance, shows this different behavior in action. In the example under discussion, the state of retardation is indicated by the decreasing and flattening trend of the crack growth rate at the end of the crack initiation period. Similar decreasing trends in small crack growth were documented by Newman et al. [38], as seen in Figure 8a at S1. In Figure 8a, the growth of PSC is represented by the dotted lines, which occur to the left of the LC growth curve, indicated by a solid line. The threshold for long crack growth is defined by the SIF range, $\Delta K_{th}$. Values of $\Delta K$ below this threshold indicate regions where no crack growth is observed.

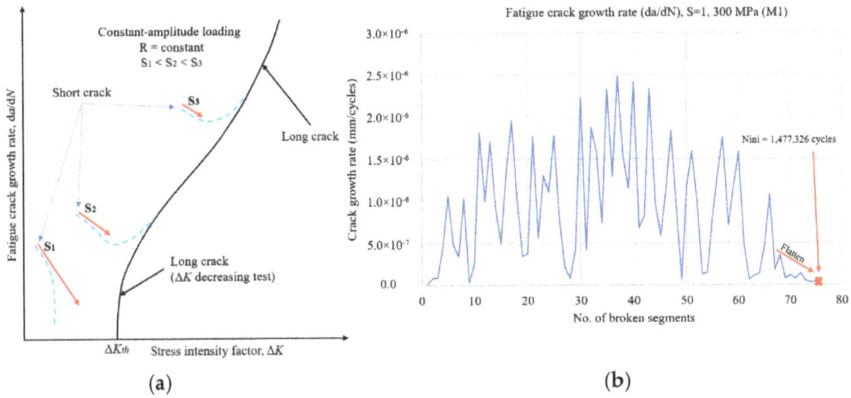


**Figure 8.** (**a**) Fatigue crack growth for rate for short and long cracks (data from Ref. [38]); (**b**) fatigue crack growth rate in case of retardation for stress level 300 MPa (M1).

As depicted in Figure 8a, the data points indicate short crack growth declines in the number of cycles and merge with the LC curve at increased loading levels (S2 and S3), as described by Newman et al. [38]. This merging is an indication that the cracks have transitioned from the short crack initiation phase to the LC growth regime. Therefore, the completion of the crack initiation process can be precisely identified at the juncture where the short crack initiation data align with the LC curve, indicating that the behavior of the crack growth has entered the LC propagation stage. The proposed idea is that the crack had changed from the crack initiation stage to the LC growth phase when the crack growth rate reduced noticeably (Figure 9c). As a result, the total number of cycles required for crack initiation could be calculated by adding the cycles required for the nucleation of all segmental cracks until a discernible crack growth rate reduction was noticed. However, at a lower loading point (S1), short cracks may stop growing and not continue propagating to an LC because of the vast number of cycles with shallow $da/dN$ magnitude. This condition causes the oscillation of the $da/dN$ to follow the usual downward trend, as shown in Figure 8b. In Figure 8b, the case study conducted at a stress amplitude of 300 MPa demonstrates a flattening trend in the $da/dN$ graph as it approaches the final data points, indicating crack retardation. The cumulative number of cycles associated with the crack initiation stage in the microstructural model can be determined from the aggregate number of cycles associated with the initiation of each segmental crack. The existence of flattened $da/dN$ curves demonstrates crack retardation and marks the end of this stage. For the exact situation shown in Figure 8b, 1,477,326 cycles were cumulated. The pattern curve in Figure 8b shows periodic fluctuations in the fatigue crack growth rate, indicating that the crack interacts variably with the material's microstructure. These interactions suggest that different regions of the grains possess a distinct resistance to crack propagation.

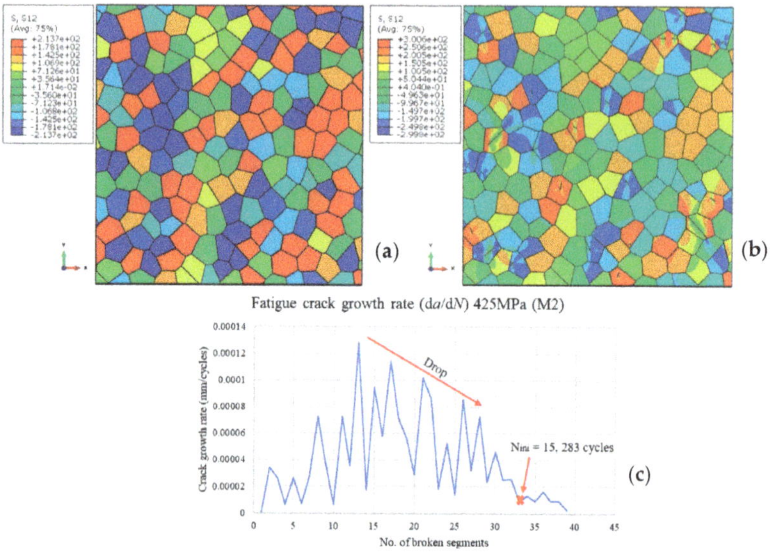


**Figure 9.** (**a**) Microstructure under a stress of 425 MPa; (**b**) microstructure depicting scattered crack initiation following 33 iterations; (**c**) fatigue crack growth rate plotted at 425 MPa stress amplitude.

The micro-model shown in Figure 9b contains a nucleated MSC, in which the shear stress distribution field can also be recognized. The projected number of cycles needed to complete initiation is shown in Figure 9c on the basis of the above conditions. In this specific instance, 15,283 cycles are required for a stress level of 425 MPa at microstructure 2 (M2). Short cracks are defined as those having a depth of less than 1 mm, as stated by Anderson [30]. Most fatigue fractures remain in the short crack phase for the bulk of their lifetime. Thus, learning about the characteristics of these flaws is of great practical importance.

The model illustrates that cracks typically nucleate in a scattered manner according to the damage course, especially in grains with favorable orientations. According to the TMM, crack formation occurs within the grains when the conditions for microcrack nucleation are favorable. Existing crack segments tend to spread along the entire grain, thus causing localized stress relaxation and concentration at their ends. This condition makes it more likely that new cracks will form in the area. The total number of cycles, $N_{ini}$, required for the entire crack initiation is equal to the cumulative sum of cycles required for all segmental cracks that form until the observed rate drop, as illustrated in Figure 9c.

As illustrated in Figure 10, a crack initiation Wöhler (S-N) curve has been generated to graphically depict the initiation phase of a fatigue crack under varied stress amplitudes. The curve is a significant improvement over the previous results in [2] due to the addition of plasticity to the global model and the microstructure models.

In the microstructure model, plastic deformation will happen when the material is stressed beyond its elastic limit, leading to crack nucleation. Considering plasticity is essential for understanding and predicting material fatigue under various high, medium, and low loads. The black curve demonstrates the average fatigue crack initiation from three different micro-models with similar grain size predictions for each data point. In the absence of experimental data of the 350, 400, and 524 MPa stress amplitudes, the TMM produced a result that aligned with the overall trend of the Wöhler curve. As seen in Table 4, more than a million cycles at a stress amplitude of 300 MPa occurred during the experiment. The average number of cycles over the three microstructures is 1,479,969, as obtained by the TMM equation, shown in Table 4.

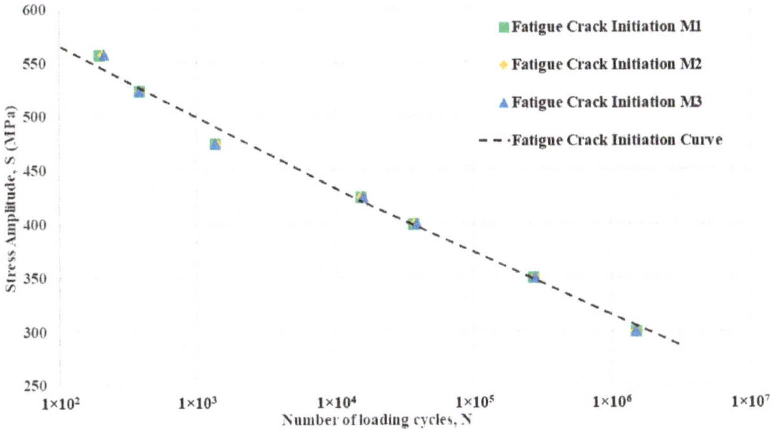


**Figure 10.** Crack initiation Wöhler (S-N) curve for fatigue crack initiation.

**Table 4.** Number of fatigue crack initiation cycles for artificial microstructures and complete fracture from the experimental observation.

| Stress Amplitude (MPa) | Number of Cycles for Fatigue Crack Initiation | | | | Number of Cycles for Complete Fracture (Experimental Data at Room Temperature) |
|---|---|---|---|---|---|
| | M1 | M2 | M3 | Average | |
| 300 | 1,477,326 | 1,479,435 | 1,480,134 | 1,479,969 | >1 million |
| 350 | 266,628 | 277,132 | 276,872 | 273,544 | Not provided |
| 400 | 36,223 | 36,937 | 38,255 | 37,138 | Not provided |
| 425 | 15,033 | 15,283 | 15,761 | 15,359 | 20,944 |
| 475 | 1360 | 1396 | 1364 | 1373 | 2361 |
| 524 | 394 | 364 | 398 | 385 | Not provided |
| 558 | 194 | 210 | 213 | 206 | 356 |

LC propagation was examined using the LEFM parameter, the SIF $K$. The model presented in Figure 7 calculated this value using the Abaqus-built XFEM. Table 5 contains a part of the $\Delta K$ values calculated for different crack lengths under a 425 MPa cyclic load. The experimental value of critical SIF 185.5 MPa$\sqrt{\text{m}}$ was provided by Materialprüfungsanstalt, Universität Stuttgart [24].

**Table 5.** Critical SIF $K_{Ic}$ computed for different crack lengths under a 425 MPa stress amplitude.

| Crack Length, a (mm) | 0.21 | 0.42 | 0.44 | 0.70 | 0.90 | 1.19 | 1.28 | 1.30 | 1.301 |
|---|---|---|---|---|---|---|---|---|---|
| $K_{Ic}$, MPa$\sqrt{\text{m}}$ | 49.93 | 74.83 | 77.07 | 104.79 | 127.66 | 166.06 | 181.61 | 184.39 | 185.84 |

As shown in Equation (4), incorporating the Paris law approach can estimate the number of cycles required for crack propagation to reach final failure. This procedure is dependent on the $\Delta K$ values determined in the previous phase. The blue line in Figure 11 is the outcome of this integration for a 425 MPa stress amplitude. In the micromechanical model, the final crack length is determined by summing up the projected lengths of individual crack segments identified in the grains. In order to accurately represent the transition from initiation to the propagation stage, the dimensions of the microstructural model were increased to 0.5 mm × 0.3 mm. Based on the micromechanical model's dimensions of 0.3 mm × 0.3 mm and the extended microstructure model's dimensions of 0.5 mm × 0.3 mm, the initial crack segment length is found to be very similar, namely, 0.21 mm and 0.19 mm, respectively. This accumulation marks the point transition from micro- to mesomechanical modeling. Within the mesomechanical scale model, the crack length begins at 0.21 mm. Table 5 shows the breakdown of the SIF at different crack lengths.

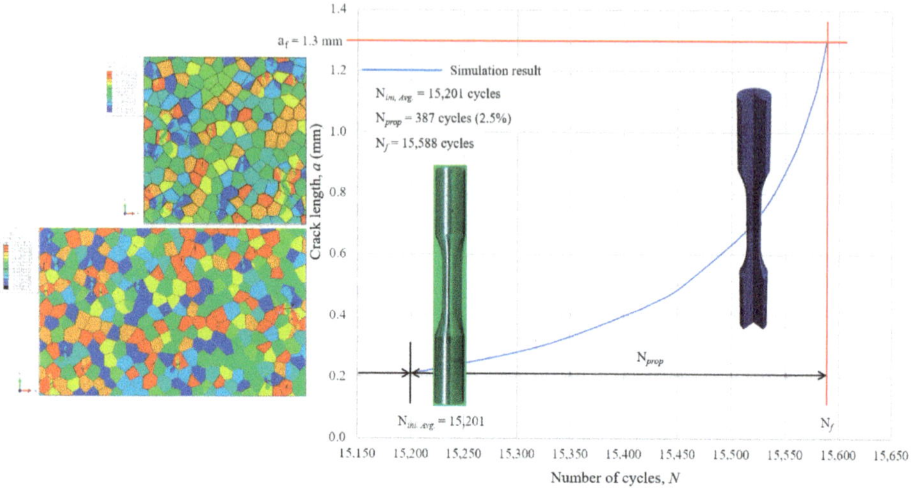


**Figure 11.** Number of cycles of LC growth at 425 MPa.

The critical SIF $K_{Ic}$ is determined to be 184.4 MPa$\sqrt{m}$ at the crack length of 1.3 mm, which is nearly close to the experimental value mentioned above. The number of cycles it would take for a crack initiation to propagate and cause failure is estimated by using the Paris law. Figure 11 depicts how this was accomplished by drawing a vertical line from the intersection of the $a - N$ curve (the blue line) and the crack length, $a_f = 1.3$ mm. The total number of cycles ($N_f$) from both simulations is further explained in Figure 11, including fatigue crack initiation ($N_{ini,Avg.}$) and propagation ($N_{prop}$).

Figure 12 shows an integrated graph exhibiting the stages of fatigue crack initiation (denoted in black curve), fatigue failure curve (depicted in a light blue curve), and the overall number of cycles resulting in fatigue failure as derived from experimental data (represented by red curve). Within statistical tolerances, a difference between 2.5% and 39.0% between the fatigue crack initiation period and propagation phase is acceptable. The apparent difference can be attributed to the stress amplitude applied at the micro and macroscale level; a higher stress amplitude has a pronounced effect on the difference of fatigue crack propagation and fatigue crack initiation. The importance of fatigue crack initiation over fatigue crack propagation in predicting the fatigue life of a material is a crucial finding of this study and has been known for a long time from the literature and for other materials. Short cracks behave very differently from longer cracks during fatigue loading. Another important detail to remember is that the crack initiation phase consumes most of the fatigue life [30]. Additional investigation into the CRSS value is significant to assess its impact on the onset of fatigue crack initiation in the P91 material.

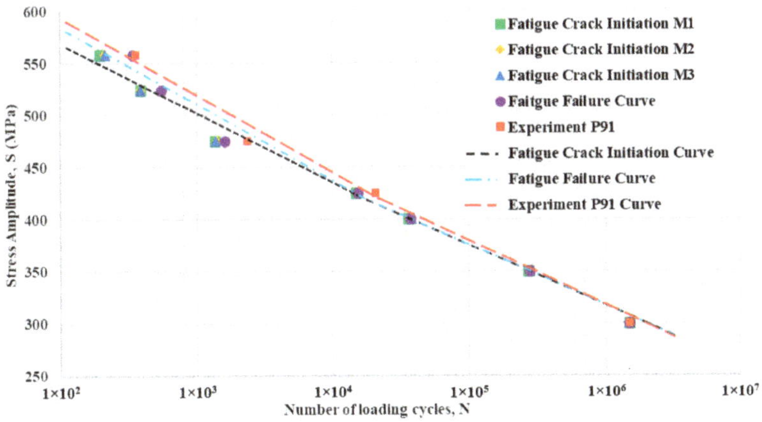


**Figure 12.** Wöhler curve for fatigue initiation and failure of X10CrMoVNb9-1 (P91).

## 4. Conclusions

On the basis of the experimental and numerical methods of estimating fatigue life using the TMM and the derivation of the Paris law, predicting fatigue life on the basis of crack initiation and propagation is promising. The following conclusions can be drawn from this work:

- The dimensions of the specimen are in accordance with ASTM E6060-04 standards.
- The local stress field within a component and the material's qualities significantly impact the initiation and development of cracks along slip planes within polycrystalline materials.
- The findings of this study are an improvement over earlier results [2] and have more realistic agreement with the Wöhler curve compared with experimental data in respect to the number of cycles. The enhancement is attributed to incorporating elastic–plastic parameters in the TMM, addressing the previously observed overestimation of cycles for short crack initiation, especially at higher stress levels in the case of neglecting additional continuum plasticity effects in the model.
- This approach offers a way to enhance the durability and dependability of polycrystalline materials in various technical applications and improve the design of their constituent parts.
- The TMM and an artificial microstructure model, in combination with VT, successfully define the fatigue crack initiation location.
- The critical SIF $K_{Ic}$ can be determined using the 2.5D axisymmetric model. Subsequently, applying the Paris law equation facilitated the calculation of the number of cycles required for fatigue crack propagation. This research emphasizes the crucial roles of the TMM and Paris law in evaluating fatigue life cycles and paves the way for the analyses of fatigue life at elevated temperatures.
- The TMM was applied to artificial microstructure models at high, medium, and low stress levels. Findings show that fatigue crack propagation affects up to 40% of the final lifetime, with a typical effect of around 15% in agreement with standard knowledge also for other materials.

**Author Contributions:** Conceptualization, M.R.B.A.R., S.S. and P.B.; methodology, M.R.B.A.R., S.S., P.B., Y.H.P.M., J.D. and T.C.; software, M.R.B.A.R., S.S. and K.J.D.; validation, M.R.B.A.R., S.S. and Y.H.P.M.; formal analysis, J.D., T.C., M.F.M. and M.I.M.A.; investigation, M.R.B.A.R., S.S., P.B., Y.H.P.M. and T.C.; resources, J.D., T.C. and M.F.M.; data curation, S.S., P.B. and Y.H.P.M.; writing—original draft preparation, M.R.B.A.R.; writing—review and editing, S.S., Y.H.P.M. and M.R.B.A.R.; visualization, M.R.B.A.R. and K.J.D.; supervision, S.S. and Y.H.P.M.; project administration, M.R.B.A.R.; funding acquisition, M.R.B.A.R. All authors have read and agreed to the published version of the manuscript.

**Funding:** This research was funded by the Public Service Department of Malaysia with reference number JPA(I)870614145415. This publication was funded by the German Research Foundation grant, "Open Access Publication Funding/2023-2024/University of Stuttgart" (512689491). Ján Dusza gratefully acknowledge the support of Alexander von Humboldt Foundation.

**Data Availability Statement:** The data presented in this study are available on request from the corresponding author.

**Conflicts of Interest:** The authors declare no conflict of interest.

## References

1. Cui, R.Y.; Hultman, N.; Edwards, M.R.; He, L.; Sen, A.; Surana, K.; McJeon, H.; Iyer, G.; Patel, P.; Yu, S.; et al. Quantifying Operational Lifetimes for Coal Power Plants Under the Paris Goals. *Nat. Commun.* **2019**, *10*, 4759. [CrossRef] [PubMed]
2. Rahim, M.R.A.; Schmauder, S.; Manurung, Y.H.P.; Binkele, P.; Ahmad, M.I.M.; Dogahe, K. Cycle Number Estimation Method on Fatigue Crack Initiation using Voronoi Tessellation and the Tanaka Mura Model. *J. Fail. Anal. Prev.* **2023**, *23*, 548–555. [CrossRef]
3. Abarkan, I.; Khamlichi, A.; Shamass, R. Low Cycle Fatigue Behavior of Circumferentially Notched Specimens Made of Modified 9Cr-1Mo Steel at Elevated Temperature. *Int. Rev. Appl. Sci. Eng.* **2022**, *13*, 54–62. [CrossRef]
4. Speicher, M.; Klenk, A.; Coleman, K. Creep-Fatigue Interactions in P91 Steel. In Proceedings of the 13th International Conference on Fracture, Beijing, China, 16–21 September 2013.
5. Saxena, A.; Narasimhachary, S. *Creep-Fatigue Crack Growth Testing of P91 Steel: Result of the Round Robin for Assessing ASTM Standard E-2760-10*; EPRI: Palo Alto, CA, USA, 2018; pp. 28–43.
6. Saad, A.A.; Hyde, C.J.; Sun, W.; Hyde, T.H. Thermal-Mechanical Fatigue Simulation of a P91 Steel in a Temperature Range of 400–600 °C. *Mater. High Temp.* **2011**, *28*, 212. [CrossRef]
7. Zhou, J.; Barrett, R.A.; Leen, S.B. A Physically-Based Method for Predicting High Temperature Fatigue Crack Initiation in P91 Welded Steel. *Int. J. Fatigue* **2021**, *153*, 106480. [CrossRef]
8. Tahmasbi, K.; Alharthi, F.; Webster, G.; Haghshenas, M. Dynamic Frequency-Dependent Fatigue Damage in Metals: A State-of-the-Art Review. *Forces Mech.* **2023**, *10*, 100167. [CrossRef]
9. Zhan, Z. Fatigue Life Calculation for TC4-TC11 Titanium Alloy Specimens Fabricated by Laser Melting Deposition. *Theor. Appl. Fract. Mech.* **2018**, *96*, 114–122. [CrossRef]
10. Huang, J.; Meng, Q.; Zhan, Z.; Hu, W.; Shen, F. Damage Mechanics-Based Approach to Studying Effects of Overload on Fatigue Life of Notched Specimens. *Int. J. Damage Mech.* **2019**, *28*, 538–565. [CrossRef]
11. Mlikota, M.; Schmauder, S. Multiscale Modelling and Simulation of Metal Fatigue and Its Applications. Ph.D. Thesis, Universität Stuttgart, Stuttgart, Germany, 2020.
12. Jezernik, N.; Kramberger, J.; Lassen, T.; Glodež, S. Numerical Modelling of Fatigue Crack Initiation and Growth of Martensitic Steels. *Fatigue Fract. Eng. Mater. Struct.* **2010**, *33*, 714–723. [CrossRef]
13. Kramberger, J.; Jezernik, N.; Göncz, P.; Glodež, S. Extension of the Tanaka-Mura Model for Fatigue Crack Initiation in Thermally Cut Martensitic Steels. *Eng. Fract. Mech.* **2010**, *77*, 2040–2050. [CrossRef]
14. Brückner-Foit, A.; Huang, X. Numerical Simulation of Micro-crack Initiation of Martensitic Steel under Fatigue Loading. *Int. J. Fatigue* **2006**, *28*, 963–971. [CrossRef]
15. Mlikota, M.; Schmauder, S.; Božić, Ž. Calculation of the Wöhler (S-N) Curve Using a Two-Scale Model. *Int. J. Fatigue* **2018**, *114*, 289–297. [CrossRef]
16. Santus, C.; Taylor, D. Physically Short Crack Propagation in Metals During High Cycle Fatigue. *Int. J. Fatigue* **2009**, *31*, 1356–1365. [CrossRef]
17. Tanaka, K.; Mura, T. A Dislocation Model for Fatigue Crack Initiation. *J. Appl. Mech.* **1981**, *48*, 97–103. [CrossRef]
18. Tanaka, K.; Mura, T. A Theory of Fatigue Crack Initiation at Inclusions. *Metall. Trans. A* **1982**, *13A*, 117–123. [CrossRef]
19. Paris, P.C.; Erdogan, F. A Critical Analysis of Crack Propagation Laws. *J. Basic Eng.* **1963**, *84*, 528–533. [CrossRef]
20. Lesiuk, G.; Smolnicki, M.; Rozumek, D.; Krechkovska, H.; Student, O.; Correia, J.; Mech, R.; Jesus, A.D. Study of the Fatigue Crack Growth in Long-Term Operated Mild Steel under Mixed-Mode (I + II, I + III) Loading Condition. *Materials* **2020**, *13*, 160. [CrossRef] [PubMed]
21. Kumar, S.; Varshney, A.; Sangal, S.; Mondal, K. Enhancement of Mechanical Properties of Modified 9Cr–1Mo (P91) Steel Using the Thermomechanical Processing and Smart Heat Treatment Protocol. *Mater. Sci. Eng. A* **2022**, *844*, 143177. [CrossRef]
22. Sakthivel, T.; Sasikala, G.; Vasudevan, M. Role of Microstructures on Heterogeneous Creep Behaviour Across P91 Steel Weld Joint Assessed by Impression Creep Testing. *Mater. Charact.* **2020**, *159*, 109988. [CrossRef]
23. Rejeesh, R.; Barik, R.K.; Mitra, R.; Kostryzhev, A.; Das, C.R.; Albert, S.K.; Chakrabarti, D. Effect of B and N Content and Austenitization Temperature on the Tensile and Impact Properties of Modified 9Cr-1Mo Steels. *Metals* **2023**, *13*, 1124. [CrossRef]
24. Speicher, M. *Versuchsdurchführung in Compliancetechnik gemäß MPAS-PA 52220-05*; Auftragsabwicklung Bruchmechanikversuch: Stuttgart, Germany, 2013; p. 1.
25. ASTM Committee E08. *ASTM E606-04: Standard Practice for Strain Controlled Fatigue Testing*; ASTM International: West Conshohocken, PA, USA, 2004; pp. 1–16.

26. Hatakeyama, T.; Sawada, K.; Suzuki, M.; Watanabe, M. Microstructure Development of Modified 9Cr-1Mo Steel During Laser Powder Bed Fusion and Heat Treatment. *Addit. Manuf.* **2023**, *61*, 103350. [CrossRef]
27. 4.2.1 Plasticity Models: General Discussion. Available online: https://classes.engineering.wustl.edu/2009/spring/mase5513/abaqus/docs/v6.6/books/stm/default.htm?startat=ch04s02ath101.html (accessed on 26 September 2023).
28. Kujawski, D.; Vasudevan, A.K.; Ricker, R.E.; Sadananda, K. On 50 Years of Fatigue Crack Closure Dispute. *Fatigue Fract. Eng. Mater. Struct.* **2023**, *46*, 2816–2829. [CrossRef]
29. Mlikota, M.; Staib, S.; Schmauder, S.; Božić, Ž. Numerical Determination of Paris Law Constants for Carbon Steel Using a Two-Scale Model. *J. Phys. Conf. Ser.* **2017**, *843*, 012042. [CrossRef]
30. Anderson, T.L. *Fracture Mechanics: Fundamental and Applications, 4th ed*; CRC Press Taylor & Francis Group: Baco Raton, FL, USA, 2017; pp. 473–476.
31. Broek, D. *The Practical Use of Fracture Mechnics*, 3rd ed.; Kluwer Academic Publishers: Dordrecht, The Netherlands, 1997; pp. 123–166.
32. Abaqus 6.13: Theory Guide. Available online: http://130.149.89.49:2080/v6.13/books/stm/default.htm (accessed on 26 September 2023).
33. Abaqus Scripting Reference Guide. Available online: http://130.149.89.49:2080/v6.13/books/ker/default.htm (accessed on 26 September 2023).
34. Yang, X.; Shi, C.; Cao, H.; Sun, S.; Liu, G. Study on Intelligent Classification of Aging Heat-Resistant Materials. *ACS Omega* **2023**, *8*, 5300–5305. [CrossRef] [PubMed]
35. Li, D.F.; Golden, B.J.; O'Dowd, N.P. Multiscale Modelling of Mechanical Response in a Martensitic Steel: A Micromechanical and Length-Scale-Dependent Framework for Precipitate Hardening. *Acta Mater.* **2014**, *80*, 445–456. [CrossRef]
36. Mlikota, M.; Schmauder, S.; Hummel, M. Modelling of Overload Effects on Fatigue Crack Initiation in Case of Carbon Steel. *Fatigue Fract. Eng. Mater. Struct.* **2017**, *40*, 1182–1190. [CrossRef]
37. Dogahe, K.J.; Guski, V.; Mlikota, M.; Schmauder, S.; Holweger, W.; Spiller, J.J.; Mayer, J.; Schwedt, A.; Görlach, B.; Wranik, J. Simulation of the Fatigue Crack Initiation in SAE 52100 Martensitic Hardened Bearing Steel During Rolling Contact. *Lubricants* **2022**, *10*, 62. [CrossRef]
38. Newman, J.C., Jr.; Phillips, E.P.; Swain, M.H. Fatigue-Life Prediction Methodology Using Small-Crack Theory. *Int. J. Fatigue* **1999**, *21*, 109–119. [CrossRef]

metals MDPI

*Article*

# Modeling the Evolution of Grain Texture during Solidification of Laser-Based Powder Bed Fusion Manufactured Alloy 625 Using a Cellular Automata Finite Element Model

**Carl Andersson and Andreas Lundbäck ***

Department of Engineering Sciences and Mathematics, Division of Solid Mechanics, Luleå University of Technology, 971 87 Luleå, Sweden; carl.andersson@ltu.se
* Correspondence: andreas.lundback@ltu.se

**Abstract:** The grain texture of the as-printed material evolves during the laser-based powder bed fusion (PBF-LB) process. The resulting mechanical properties are dependent on the obtained grain texture and the properties vary depending on the chosen process parameters such as scan velocity and laser power. A coupled 2D Cellular Automata and Finite Element model (2D CA-FE) is developed to predict the evolution of the grain texture during solidification of the nickel-based superalloy 625 produced by PBF-LB. The FE model predicts the temperature history of the build, and the CA model makes predictions of nucleation and grain growth based on the temperature history. The 2D CA-FE model captures the solidification behavior observed in PBF-LB such as competitive grain growth plus equiaxed and columnar grain growth. Three different nucleation densities for heterogeneous nucleation were studied, $1 \times 10^{11}$, $3 \times 10^{11}$, and $5 \times 10^{11}$. It was found that the nucleation density $3 \times 10^{11}$ gave the best result compared to existing EBSD data in the literature. With the selected nucleation density, the aspect ratio and grain size distribution of the simulated grain texture also agrees well with the observed textures from EBSD in the literature.

**Keywords:** additive manufacturing; SLM; microstructure; equiaxed; columnar; CET; numerical; simulation

## 1. Introduction

Additive manufacturing (AM) is transforming the manufacturing industry by allowing complex-shaped geometries to be manufactured while reducing material waste compared to subtractive methods such as milling [1]. In additive manufacturing, the material is added in a layer-by-layer approach and by subsequently melting the added material with a heat source it is fused into a solid structure [2]. Laser-based powder bed fusion (PBF-LB) is one of the most common additive manufacturing methods where a thin layer of metal powder is first spread and then selectively melted using a laser beam, also known as selective laser melting (SLM) [3]. The nickel-based superalloy 625 is a common AM material and it is widely used in the aerospace, chemical, and petrochemical industries due to its good yield strength, creep strength, and weldability [4], as well as its good oxidation and corrosion resistance [5]. Dendritic columnar and equiaxed microstructures are typically seen in alloy 625 parts produced by PBF-LB [6,7]. These microstructural features arise from dendritic crystal growth. As the crystals grow, grains are forming consisting of dendritic crystals that share the same crystallographic orientation. The grain texture refers to the distribution of these grains and their crystallographic orientations. Each oriented crystal has its preferential growth direction. If the preferential growth direction is aligned with the maximum heat flow direction it is more likely to sustain its growth and outcompete other growing crystals. This is known as competitive grain growth which has a significant influence on the grain textures of PBF-LB produced materials. The maximum heat flow direction in additive manufacturing is along the build direction which causes pronounced columnar grain textures oriented along the build direction as seen in Malmelöv et al. [8]. The mechanical properties of the produced material will vary depending on the obtained grain texture. Through modeling, the desired mechanical properties can be obtained, while costly, time consuming, and material-demanding physical tests can be avoided. To be able to model the grain texture evolution, the solidification process in AM needs to be understood. Figure 1 illustrates the solidification process and the characteristics of the as-solidified grain texture. Additive manufacturing is comparable to many repetitive welds and therefore it is assumed that the nucleation and grain growth mechanisms are similar to those of the welding described by Kou [9]. There are two competing nucleation mechanisms that take place while the melt pool is solidifying: epitaxial nucleation and heterogeneous nucleation. Epitaxial nucleation occurs [10] at the interface between solid material and the end of the solidification zone. Pre-existing crystals that are partially re-melted start to grow into the undercooled liquid to form columnar grains. Heterogeneous nucleation occurs in the undercooled liquid where grains nucleate due to bulk nucleation which forms equiaxed grains. If the equiaxed zone grows large enough, the columnar front will be prevented from growing further and the Columnar-to-Equiaxed Transition (CET) takes place. By varying the process parameters, the equiaxed zone can be more likely to sustain its growth due to more favorable conditions which promotes the CET [11,12].

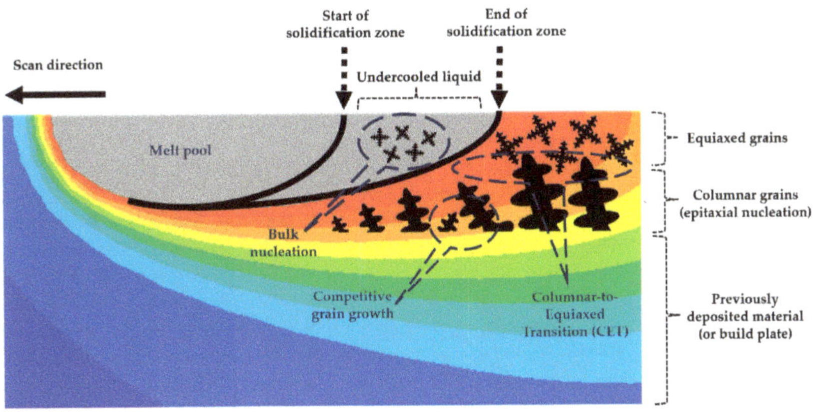


**Figure 1.** Nucleation and grain growth mechanisms in welding [9] which is assumed to be similar as in additive manufacturing.

There are three common models used to predict the grain texture during solidification: the Phase Field (PF) model, the Cellular-Automata (CA) model, and the Kinetic Monte-Carlo (KMC) model [13,14]. The PF model is used to predict the dendritic growth during solidification, resulting in a high-resolution prediction of the dendritic growth but is thus also limited to predictions containing a few dendrites [15]. The CA model describes the dendritic growth on a cell grid where each cell can exist in different states and the cells evolve between states according to predefined transition rules. The model applies to mesoscale objects since only the growth of the dendrite tip is considered which significantly reduces the computational time compared to the PF model that considers the entire liquid–solid interface of the dendrite. The KMC model is computationally efficient enough to reproduce full 3D microstructures that agrees with experimental findings [16], but it cannot predict grain texture since it does not incorporate the effects of crystallographic orientation and competitive grain growth [13,16]. The CA model was deemed as most suitable to predict the characteristics of the printed grain texture efficiently and accurately in a mesoscale context applied to PBF-LB.

In the 1990s, Gandin and Rappaz developed a CA model to predict the grain texture evolution during solidification of casted parts [12,17,18]. CA models have gained popularity for simulating the grain texture evolution in alloys manufactured by PBF-LB due to the advancements in the field of additive manufacturing. Typically, 2D CA models are used for their computational efficiency while being capable of simulating realistic grain textures [19,20]. Dezfoli et al. [19] used a 2D CA model to simulate the grain texture from single-track PBF-LB processed alloy 718. A cell spacing of 0.2 μm was used in their CA model when considering the PBF-LB process conditions of 150 W for three different scan velocities of 500, 400, and 300 mm/s. The cell spacing is an important parameter that controls the spatial resolution in the CA model. Rai et al. [20] developed a 2D CA model to predict the grain texture of alloy 718 consisting of 10 layers processed by electron beam powder bed fusion (PBF-EB). The CA model with a cell spacing of 0.5 μm was used to compute the grain texture based on the thermal history obtained with a 2D Lattice Boltzmann model. In their case, the process condition for PBF-EB was a scan velocity of 2200 mm/s and a power of 594 W.

Although 2D models do produce realistic results for AM, they do not capture all aspects of grain growth. For example, the 2D model would not be able to capture the influence on the grain texture when grains are growing out of the plane. Numerous researchers have developed 3D CA models capable of simulating the grain texture evolution that also captures the growth of grains out of the plane [21,22]. Koepf et al. [21] developed a 3D CA-FE model that re-uses the thermal field obtained using a finite element (FE) model with four printed layers. They predicted the grain texture of the nickel-based alloy CMSX-4 produced by PBF-EB. With the re-use of the temperature field, the computational time was significantly reduced. The process conditions included a power of 300 W and a scan velocity of 500 mm/s. A spatial resolution of 10 μm was used to capture the evolution of the grain texture. Teferra and Rowenhorst [22] developed a 3D CA model where parallelization was implemented to increase computational efficiency. The PBF-LB process with a power of 175 W and a scan velocity of 500 mm/s was simulated to study the grain texture evolution of 316 L stainless steel using a spatial resolution of 1.875 μm in the CA model.

First and foremost, in the highlighted findings mentioned above, it is seen that the spatial resolution is lower for 3D models than for 2D models, and thus the accuracy of a 3D model is inherently lower. Hence, there is a need for 2D CA models that can be used both with a fine spatial and temporal resolution while also giving a reasonable computational time. In 2D CA modeling, the two competing nucleation mechanisms have a significant influence on the predicted grain texture. One of these is, as mentioned earlier, heterogeneous nucleation, and the choice of nucleation parameters are essential for success in grain texture modeling as highlighted in the review by Körner et al. [13]. The importance of choosing nucleation parameters for 2D CA models is often neglected such that nucleation density and heterogeneous nucleation are not considered. In this work, a method of finding suitable nucleation parameters for heterogeneous nucleation is proposed by calibrating them with respect to existing EBSD data in the literature. Secondly, the largest contribution to the computational time is from the thermal model. Therefore, there is a need for efficient 3D FE models that can resolve the temperature history for complex scan strategies seen in PBF-LB. In this work, a 3D FE heat transfer model is developed capable of simulating the temperature history over many deposited layers. Thermal models coupled with fluid flow such as in Dezfoli et al. [19] are used in conjunction with models for grain texture evolution. These models have the possibility to give a high resolution of the temperature field but with the price of a high computational cost. Typically, 2D Lattice Boltzmann methods are also used coupled with particle interactions such as in Rai et al. [20]. However, since heat transfer is a 3D phenomenon, the accuracy of the temperature history from such a model will be inherently lower and it cannot resolve the temperature history involving rotation scanning strategies which are often used in PBF-LB.

The aim of this work is to develop an efficient 2D CA-FE model that can predict the grain texture of the nickel-based superalloy 625 processed by PBF-LB. Both heterogeneous nucleation and epitaxial nucleation are considered in the model. Such a model can be used to obtain a better understanding of how the grain texture of the printed material is influenced by the many process parameters in PBF-LB. The prediction is based on the temperature history obtained with a 3D FE heat transfer model for a set of process parameters from the work by Malmelöv et al. [8]. The heat transfer model combined with a 2D CA model is proposed to efficiently compute grain growth and nucleation in alloys produced by PBF-LB. This enables an efficient modeling approach capable of predicting the grain texture in PBF-LB which evolves over many deposited layers. Nucleation parameters are determined through calibration with regard to existing EBSD data in the literature. The PBF-LB process conditions are then changed relative to the calibration case for model validation to see if the model captures changes in the grain texture depending on the process conditions.

## 2. Materials and Methods

This section describes the coupled CA-FE model for simulating the grain texture of PBF-LB processed alloy 625. The coupling between the CA and FE model is a one-way coupling. The temperature field is first computed with the FE model and the grain texture is computed as a post-process procedure after all timesteps are computed in the FE model. The finite element model is simulated in MSC Marc with in-house developed AM subroutines and the temperature history is imported into MATLAB where the computations of the CA model are made. The grain texture is computed from a two-dimensional cross section based on the thermal history obtained from the finite element model.

### 2.1. Finite Element Model

The finite element model was developed to take into account the process conditions for PBF-LB processed alloy 625 as presented in Malmelöv et al. [8]. A wall geometry (length 60 mm, width 4 mm, and height 30 mm) was built in their work and the thermal history for a small part of that build is simulated here (total hatch width 540 μm, scan length 300 μm, total build height 200 μm). The FE model consists of 15 deposited layers with a layer thickness of 20 μm where each layer consists of 9 hatches with the hatch spacing 60 μm and hatch length 300 μm. The scanning strategy was bi-directional and the elements were activated according to the "inactive" element approach. Figure 2 illustrates the multi-layer deposition model composed of the finite element discretization of all deposited layers and the build plate.

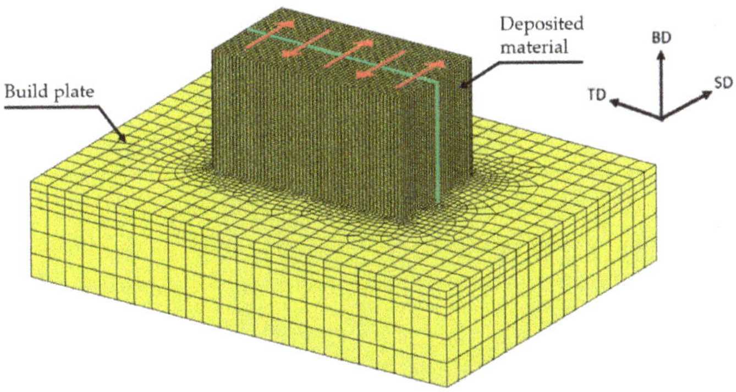


**Figure 2.** Finite element discretization of the deposited material and substrate where the grain texture is predicted with the CA model in the cross-section in the middle of the deposited layers (green lines). The red arrows indicate the bi-directional scanning strategy.

The discretization of the added layers consists of hexagonal elements with a height of 5 μm which means that one layer consists of four elements in the build direction. In between the added hatches and layers there is a short time of cooling to take a real-life scenario into account. Cooling between hatches was set to the total width of the printed wall divided by the scanning velocity. Cooling between layers was set to the total number of hatches of the printed wall multiplied by the time for cooling between hatches plus additional time for powder recoating. The grain texture is modelled on the BD-TD plane (build direction, transverse direction) indicated by the green lines in the figure. Two different process parameters are considered, case A and case B, where the power and scan velocity changes. Case A is used for model calibration and case B is used for model validation of the simulated grain texture.

The temperature history is computed from the finite element model by solving the heat conduction equation,

$$\frac{\partial}{\partial x}\left(k\frac{\partial T}{\partial x}\right)+\frac{\partial}{\partial y}\left(k\frac{\partial T}{\partial y}\right)+\frac{\partial}{\partial z}\left(k\frac{\partial T}{\partial z}\right)+\dot{q}=\rho c_p\frac{\partial T}{\partial t} \quad (1)$$

where $T$ is the temperature, $\dot{q}$ is the internal heat generation, $\rho$ is the density, and $t$ is the time. The heat conductivity $k$ and specific heat $c_p$ is given as a function of the temperature as shown in Figure 3.

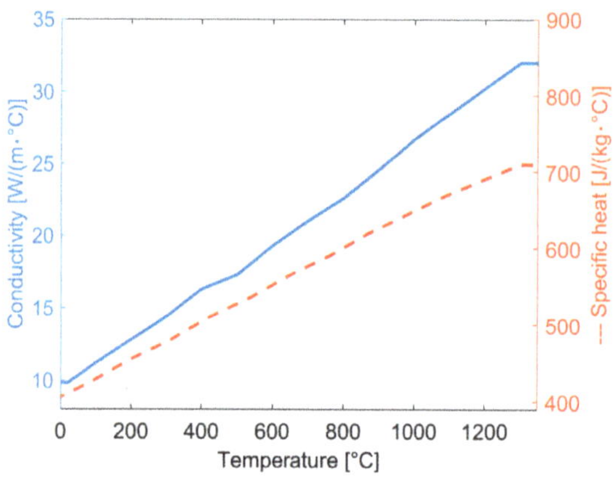


**Figure 3.** Conductivity [23] and specific heat [24] for alloy 625.

The initial condition is applied as,

$$T(x,y,z,0)=T_0 \quad (2)$$

where $T_0$ is the ambient temperature, and the boundary conditions for a free surface are applied as,

$$k\frac{\partial T}{\partial n}-q_s+h(T-T_0)+\sigma\varepsilon(T^4-T_0^4)=0 \quad (3)$$

where $\frac{\partial T}{\partial n}$ is the temperature change with respect to the normal direction of the surface and $q_s$ is the external heat flux. $h(T-T_0)$ is heat flow from convection where $h$ is the heat transfer coefficient. The term $\sigma\varepsilon(T^4-T_0^4)$ gives the heat dissipation from radiation where $\sigma$ is Stefan-Boltzmann's constant and $\varepsilon$ is the emissivity. Dissipation of heat through convection and radiation is considered during deposition and cooling between layers on the top surfaces with the emissivity and convective heat transfer coefficient set to 0.28 and 18 $W/(m^2{\cdot}K)$, respectively [25]. After cooling between layers, recoating occurs and another powder layer is spread onto the deposited layer. During this time, the convection and radiation is inactivated in the model. This is motivated by the fact that the conductivity of the powder is considerably much lower than that of the solidified material [26]. A heat flux film of 500 $W/(m^2{\cdot}K)$ is applied as a boundary condition to the bottom and the sides of the build plate to represent the heat that is conducted to the remaining parts of the build plate and to the fixture of the build plate [8]. Latent heat was set to 150 kJ/kg in the interval 1189–1336 °C for both solidification and melting [27]. The latent heat is released as a linear function between the liquidus and solidus temperature.

Goldak's double ellipsoid heat input model [28] is used to replace the physics of the generation of heat and to mimic the heat input from the laser beam. This model applies a heat flux to all integration points inside of the ellipsoidal geometry which can be described by the following equation,

$$q_{f,r}(x,y,z,t) = \frac{\beta 6\sqrt{3} f_{f,r} Q}{abc_{f,r}\pi\sqrt{\pi}} e^{\frac{-3x^2}{a^2}} e^{\frac{-3y^2}{b^2}} e^{\frac{-3(z+v(\tau-t))^2}{c_{f,r}^2}} \tag{4}$$

where $Q$ is the power input, $v$ is the velocity of the heat source, $t$ is the time, $\beta$ is a scale factor, $\tau$ defines the position of the heat source at $t = 0$, and $f_{f,r}$ defines the fraction of heat deposited in the forward or rear region. The subscripts $f$ and $r$ denotes the forward and rear part of the ellipsoid. The lengths $a$, $b$, and $c$ are half of the width, depth, and length of the ellipsoid. The net input power $Q$ is given by,

$$Q = \eta Q_{nom} \tag{5}$$

where $Q_{nom}$ is the nominal heat input and $\eta$ is the efficiency factor accounting for losses such as reflection and spattering. Table 1 gives a summary of the parameters used in the finite element model. The heat source parameters were chosen based on previous experience and it was ensured that the entire added layer was melted and that the previous deposited layers and hatches were partially remelted.

**Table 1.** Parameters in the finite element model.

| Property | Value | Unit |
|---|---|---|
| Power input $Q_A$ [8] | 100 | W |
| Scan velocity $v_A$ [8] | 1200 | mm/s |
| Power input $Q_B$ [8] | 160 | W |
| Scan velocity $v_B$ [8] | 1400 | mm/s |
| Heat source efficiency factor $\eta$ | 0.34 | - |
| Half width of heat source $a$ | 35 | µm |
| Depth of heat source $b$ | 20 | µm |
| Forward length of heat source $c_f$ | 15 | µm |
| Rear length of heat input source $c_r$ | 25 | µm |
| Initial and ambient temperature $T_0$ | 500 | °C |
| Density $\rho$ [24] | 8440 | kg/m$^3$ |

To ensure that the correct amount of heat input is added to the model in each timestep, the scale factor $\beta$ is added to Equation (4). This scale factor has two parts that it compensates for. The first part is related to the 5% cut-off limit that was proposed by Goldak et al. [28]. This cut-off limit is such that if the volumetric heat flux is below 5% of the peak value it is set to zero. In Lindgren [29], an analytical integration of the heat source equation with the given boundary conditions shows that only 89% of the power is supplied into the model. The second part is related to the loss of heat input due to the discretization. A coarser mesh will generally give a larger loss from the numerical integration of the heat source equation. This loss will vary between each time step depending on the element length in the welding direction, the velocity, and the time step length. The scale factor $\beta$ is added to Equation (4) to compensate for the heat loss. The scale factor is computed as the ratio between the wanted heat input over the actual heat input. Since the second part of the heat input loss varies in a non-predictive way, it must be computed in each time step. In the current approach, it is conducted such that the scale factor from the previous time step is used as a first guess in the first iteration of the current time step. The scale factor is then computed and applied in the subsequent iteration. Principally, the change in the scale factor can perturbate the solution between the first and second iterations. However, in practice it has been seen that there is little to no increase in the total number of iterations when solving an AM model using this approach.

### 2.2. 2D Cellular Automata Model

The cellular automata model is based on the framework developed by Gandin and Rappaz [12,17,18] and is described in this section. The temperature history is imported from the finite element model into MATLAB and a linear interpolation to the CA cell grid is made. A finer grid was used for the CA model because it is not as computationally expensive as the FE model. The CA grid contains cells with the spacing $l$ between cells and the spacing between cells was set to 0.5 μm compared to the 5 μm element size in the FE model. The temporal resolution is also refined in the CA model according to,

$$\partial t_{CA} = 0.25 \cdot \frac{l}{\max(v(\Delta T))} \quad (6)$$

where $v(\Delta T)$ is the growth velocity of the dendrite tip as a function of the undercooling. This adaptive timestep algorithm ensures that the dendritic network does not grow more than a quarter of the cell spacing in a single timestep.

Each cell in the grid has its own defined neighborhood and the cell can exist in a solid, nucleated, growing, or liquid state. The neighborhood controls the transition between the states and a Moore's second-order neighborhood with 24 neighbors was used. The cell is said to be liquid when its temperature $T$ is higher than the liquidus temperature $T_L$ and the nucleation of a liquid cell is controlled by its undercooling $\Delta T$ defined as the change in temperature below the liquidus temperature. From the liquid phase, a cell can be nucleated into the growing phase and the growth is computed according to a modified version of the decentered square growth algorithm developed by Gandin and Rappaz [17]. The cell becomes solid when the temperature is below the solidus temperature $T_S$.

#### 2.2.1. Nucleation

Additive manufacturing has many similarities to welding and it is assumed that the nucleation mechanisms are similar to those of welding. Kou [9] notes that one of the main types of nucleation in welding is epitaxial nucleation where pre-existing grains at the fusion boundary are partially remelted which reactivates the grain growth without altering the crystallographic orientation. This type of nucleation is considered in the model. Non-epitaxial growth at the fusion boundary is not considered since this is an effect when two different materials (or different crystal structures) are joined together. This is an important nucleation mechanism for aluminum alloys, and it was considered in the CA model by Mohebbi and Ploshikhin [30]. In this model, both the substrate and the deposited material are alloy 625 where epitaxial growth largely influences the grain evolution in PBF-LB [7] and non-epitaxial nucleation at the fusion boundary is neglected. New grains can also form from the nucleus caused by dendrite fragmentation, grain detachment, surface nucleation, and heterogeneous nucleation [9]. Dendrite fragmentation and grain detachment occur when convection carries detached dendrite fragments or partially detached grains into the weld pool which forms new nuclei. Surface nucleation can occur when a stream of cooling gas causes undercooling on the surface inducing nucleation effects. The effect of these three types of nucleation is neglected in the model since they are assumed to be limited in the protected environment in PBF-LB. It is also indirectly accounted for in the fourth mechanism where new nuclei originate from heterogeneous nucleation on foreign particles in the liquid melt pool.

Heterogeneous nucleation is accounted for in the model in a similar way as proposed by Gandin and Rappaz [12,18]. In their work, the nucleation was treated by heterogeneous nucleation from surfaces or through bulk nucleation in the undercooled liquid. They added the surface nucleation phenomena to capture nucleation sites on the mold walls where nuclei more easily can form due to impurities in the wall. This is not necessary for AM simulations due to the lack of surfaces such as mold walls. Heterogeneous bulk nucleation is dealt with by introducing a nucleation density parameter $n(\Delta T)$ that describes the number of nucleation sites per cubic volume. The total number of nucleation sites is then given as,

$$N_{max} = n_{max} V \tag{7}$$

where $V$ is the volume of the undercooled liquid. The rate of change for the nucleation density is given by a Gaussian distribution, hence, we can formulate the nucleation density as a function of the undercooling with the integral of the Gaussian probability density function,

$$n(\Delta T) = \int_0^{\Delta T} \frac{dn(\Delta T)}{d(\Delta T)} d(\Delta T) = \frac{n_{max}}{\Delta T_\sigma \sqrt{2\pi}} \int_0^{\Delta T} \exp(-\frac{1}{2} \frac{(\Delta T - \Delta T_{avg})^2}{(\Delta T_\sigma)^2}) \, d(\Delta T) \tag{8}$$

where $\Delta T_{avg}$ is the average undercooling and $\Delta T_\sigma$ is the standard deviation for the Gaussian distribution, respectively.

Bulk and surface nucleation is sufficient for simulating the solidification of casted parts. For additive manufacturing or any other welding procedure, the epitaxial grain growth must be considered which occurs at the fusion boundary. The nucleation for epitaxial growth is treated in a similar way as the heterogeneous nucleation by re-activating the growth of solid cells according to a Gaussian distribution at high undercooling and at temperatures near the solidus temperature. Figure 4 schematically illustrates the concept of heterogeneous nucleation for bulk nucleation and epitaxial growth. The increase in nucleation sites with respect to the undercooling is given by the blue curve, whereas the red curves describe the total number of available nucleation sites.

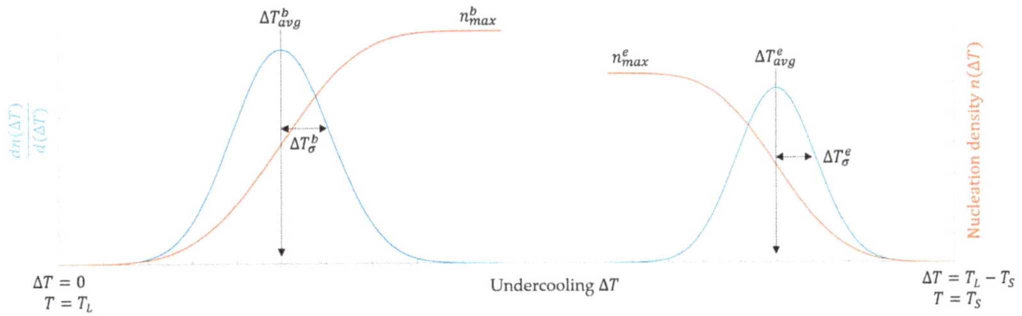


**Figure 4.** Heterogeneous bulk nucleation. The upper subscript b describes bulk nucleation, while the upper subscript e describes epitaxial growth.

In the CA model, the heterogeneous nucleation is treated by first computing the total number of nucleation sites according to Equation (7). The cross-section area is used instead of the volume for 2D CA models for a given maximum nucleation density. With the number of nuclei known, random cells are chosen and selected as nucleation sites. Each selected cell is appointed a critical undercooling $\Delta T_{crit}$, randomly chosen from the Gaussian distribution with the average $\Delta T_{avg}$ and the standard deviation $\Delta T_\sigma$. When the critical undercooling is exceeded, the nominated cell nucleates, and the cell is assigned a crystallographic orientation. Figure 5 illustrates the concept of heterogeneous nucleation implemented in a cellular automata application. All cells are nominated for epitaxial nucleation, and the reactivation of the growth occurs when the temperature is above the solidus temperature plus the contribution from the Gaussian distribution. The average of the distribution for the epitaxial nucleation was chosen such that the undercooling from the average to the solidus temperature is the same as the undercooling from the average of the heterogeneous nucleation distribution to the liquidus temperature. The standard deviation for the distribution was kept the same for both distributions. A summary of the nucleation parameters is given in Table 2.

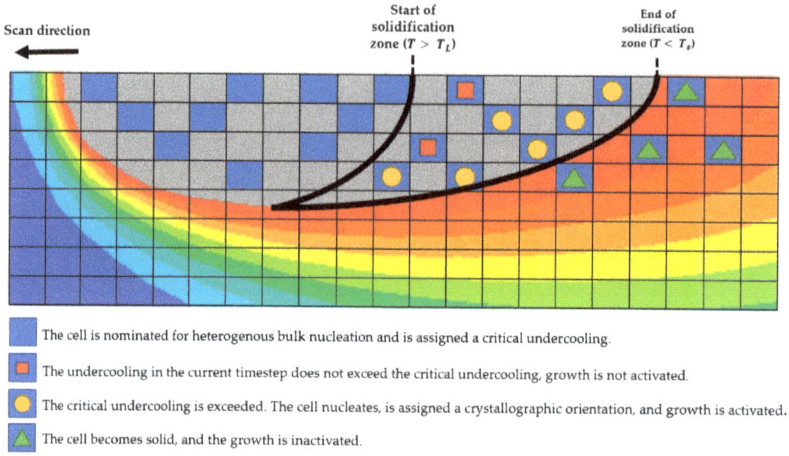


**Figure 5.** Heterogeneous nucleation in the CA model.

**Table 2.** Nucleation parameters in the CA model.

| Property | Value [°C] |
|---|---|
| Average bulk nucleation temperature $\Delta T^b_{avg}$ [18] | 5.5 |
| Standard deviation bulk nucleation $\Delta T^b_\sigma$ [12] | 0.1 |
| Average epitaxial nucleation temperature $\Delta T^e_{avg}$ | 141.5 |
| Standard deviation epitaxial nucleation $\Delta T^e_\sigma$ | 0.1 |
| Liquidus temperature $T_L$ [8] | 1336 |
| Solidus temperature $T_S$ [8] | 1189 |

### 2.2.2. Grain Growth

After nucleation, the cell is given a random crystallographic orientation $\theta$ represented by a random integer between 0 and 45 degrees. As a cell has been nucleated, it is allowed to grow. Cells with a temperature between the liquidus and the solidus temperature in the mushy zone are said to be growing. In the growing phase, the dendrite tips grow with the velocity $v(\Delta T)$. The dendrite tip velocity can analytically be computed with the Lipton-Glicksman-Kurz (LGK) model [31] and the Kurz-Giovanola-Trivedi (KGT) model [32]. A simplified expression of the LGK model is commonly used for CA models [22,33,34] which also is used in this work. This expression relates the growth velocity of the dendrite tip as a function of the undercooling as,

$$v(\Delta T) = \frac{D_L}{5.51\pi^2(-m_L(1-k))^{1.5}\Gamma}\left(\frac{\Delta T^{2.5}}{C_0^{1.5}}\right) \tag{9}$$

where $D_L$ is the diffusion coefficient in liquid, $m_L$ is the liquidus slope, $k$ is the partition coefficient, $\Gamma$ is the Gibbs-Thomson coefficient, and $C_0$ is the initial concentration. A polynomial function was fitted to this model to further simplify the calculations,

$$v(\Delta T) = a_1\Delta T + a_2\Delta T^2 + a_3\Delta T^3 \tag{10}$$

where $a_1$, $a_2$, and $a_3$ are the polynomial coefficients. The values for the growth kinetic parameters are listed in Table 3 together with the polynomial coefficients. The growth kinetic parameters are taken for a binary approximation of a nickel-based superalloy, and it is assumed that most nickel-based superalloys share similar growth kinetics.

**Table 3.** Growth kinetics parameters for a binary approximation of a nickel-based superalloy.

| Property | Value | Unit |
|---|---|---|
| Diffusion coefficient $D_L$ [35] | $3 \times 10^{-9}$ | $m^2/s$ |
| Liquidus slope $m_L$ [35] | $-10.9$ | K/wt% |
| Partition coefficient $k$ [35] | 0.48 | - |
| Gibbs – Thomson coefficient $\Gamma$ [35] | $10^{-7}$ | K·m |
| Initial concentration $C_0$ [35] | 4.85 | wt% |
| 1st polynomial coefficient $a_1$ | $-4.6 \times 10^{-4}$ | - |
| 2nd polynomial coefficient $a_2$ | $2.80 \times 10^{-5}$ | - |
| 3rd polynomial coefficient $a_3$ | $1.46 \times 10^{-7}$ | - |

In the growing stage, each cell is represented by its growing square oriented with the crystallographic orientation $\theta$ where the corners of the square represent the primary dendritic tips that grow according to Equation (10). The square envelope is visualized in Figure 6 and half of the square's side can be computed as follows:

$$L(t) = \frac{1}{\sqrt{2}} \int v(\Delta T)\, dt \qquad (11)$$

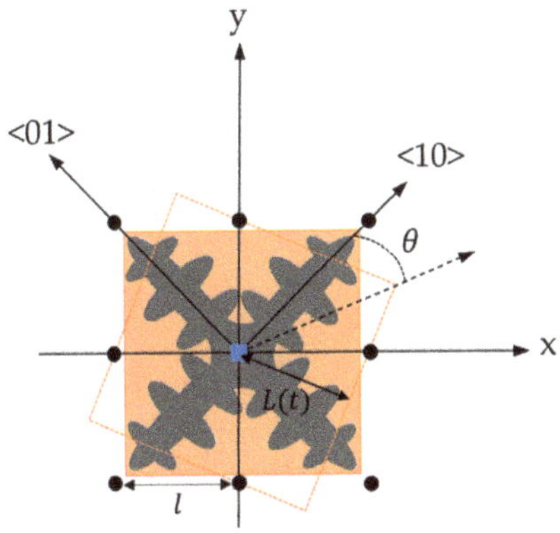


**Figure 6.** Square envelope of the growing central cell marked by a blue square and its first eight neighbors marked with black dots. The dashed square represents a cell with the crystallographic orientation $\theta$. The corners of the solid square represent the dendritic tip of a crystal misoriented by $\theta = 0$ degrees with the preferential growth vectors <10> and <01>.

The growing central cell is said to capture its neighbors when the neighboring cells are located on or inside the square envelope of the growing central cell. When the neighboring cells are captured, the captured cells inherit the crystallographic orientation $\theta$ from the central cell, and new square envelopes are initialized at each of the captured cells with the inherited orientation. The growth center of the newly captured squares is decentered from their local cell position and a virtual growth center of these cells is set at the nearest dendrite tip of the central cell, represented by the corner of the square. The cell grows according to its local undercooling and not with the undercooling at the virtual growth center. This is a similar growth algorithm as the decentered square algorithm proposed by Gandin and Rappaz [17]. It is a somewhat simplified approach since the virtual growth center is directly chosen at the nearest corner with no initial size. More importantly, this method still maintains the dendrite growth front since during the growth of the central cell, the four nearest neighbors are captured first, meaning that these captured cells will have grown to some arbitrary size before the next four nearest neighbors are captured. The size of a square is truncated to a maximum value of $\sqrt{2} \cdot l$ to allow newly initialized squares to grow to an arbitrary size before the growth of the central cell is stopped.

The growth algorithm is illustrated in Figure 7, showing how the square envelopes evolve over several growth steps. Figure 7a shows the growth of the square envelope of a single central cell (red square), having reached a size big enough to capture its four nearest neighbors. New squares are initialized on captured cells but decentered from the center of the captured cells, and the initialized squares are positioned to coincide with the nearest vertex of the cell it was captured by. The central cell and the newly initialized cells continue to grow, reaching the stage in Figure 7b where four additional cells are captured by the central cell. Four new squares are then initialized decentered from their origin and placed on one of the vertices of the central square. After this stage, the complete surrounding of the central cell has been captured and the growth of the central cell is stopped to prevent it from capturing cells that would have been captured by other cells. In the next stage, Figure 7c, the newly initialized envelopes have grown and consequently captured one of its nearest neighbors which is incorporated into the grain. A summary of the transition rules is given in Table 4.

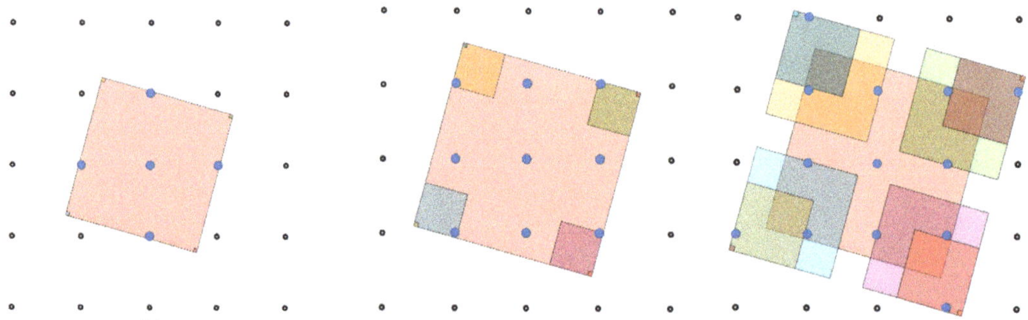


**Figure 7.** Growth over several stages for a single grain with a 15° misorientation consisting of (**a**) four cells, (**b**) nine cells, and (**c**) thirteen cells. The blue solid circles indicate the cells incorporated in the grain at each growth stage, and black hollow circles are the remaining liquid cells in the neighborhood.

**Table 4.** Summary of transformation conditions between states in the CA model.

| Cell State | Condition | Effect in CA Model |
|---|---|---|
| Liquid | Temperature $T$ above the liquidus temperature $T_L$. | The cell does not grow and can be captured by other cells. If the cell is nominated for nucleation, the cell is assigned a critical undercooling $\Delta T_{crit}$ that must be exceeded for a nucleus to form. |
| Nucleated | Undercooling $\Delta T$ above the assigned critical undercooling $\Delta T_{crit}$. | The cell can grow and is assigned a random crystallographic orientation $\theta$. |
| Growing | Temperature above solidus temperature $T_S$ and below the liquidus temperature $T_L$. | The growth of the cell's square size $L(t)$ is computed according to Equations (10) and (11). The cell can capture cells without a crystallographic orientation $\theta$, and the cell transfers the crystallographic orientation $\theta$ to captured cells |
| Solid | Temperature $T$ below the solidus temperature $T_S$. | The cell has a crystallographic orientation $\theta$, does not grow, and cannot be captured by other cells. |

## 3. Results and Discussion

In the results and discussion section, the characteristics of the simulated grain texture are first discussed. The influence of the number of nucleation sites on the grain texture is then studied and the simulated grain texture from case A is compared to existing EBSD imaging in the literature to determine a suitable nucleation density. The simulated grain texture is then compared to case B where the process conditions for PBF-LB are changed.

### *3.1. Characteristics of the As-Solidified Grain Texture*

Figure 8 illustrates the characteristics of the as-solidified grain texture where some features are highlighted and discussed. The grain texture of the build plate was first initialized by heterogeneous nucleation during uniform cooling, giving the equiaxed zone at the bottom of the figure. The 15 layers of materials were then deposited. The melt pool geometry during deposition is outlined in the figure. In the outlined melt pool geometry, it is seen that the coarse grains form in the overlapping of the melt pools. This was also seen by Andreau et al. [10] in EBSD imaging where the melt pool geometry was also computed. They reported on the finding that columnar grains grow in the overlapping of adjacent melt pools which is also seen in this case. Phase-field simulations coupled with a thermal-fluid model have also confirmed this grain growth behavior in the work by Yang et al. [36] where they saw coarse grains in the overlapping zone and elongated grains in the center of the melt pool. It is seen in Figure 8 that the elongated grains with a high aspect ratio also coincide with the center of the melt pools. It is seen that these elongated grains tend to grow from the overlapping to the center of the melt pools, creating a boundary between larger columnar grains. It is also seen that when the growth of grains is suppressed, the grains tend to obtain a V-shaped form. This was previously seen in EBSD imaging by Wang et al. [37] where they discussed that the V-shaped grain starts at the overlap between two adjacent melt pools in a lower layer and ends at the overlap between two adjacent melt pools in an upper layer.

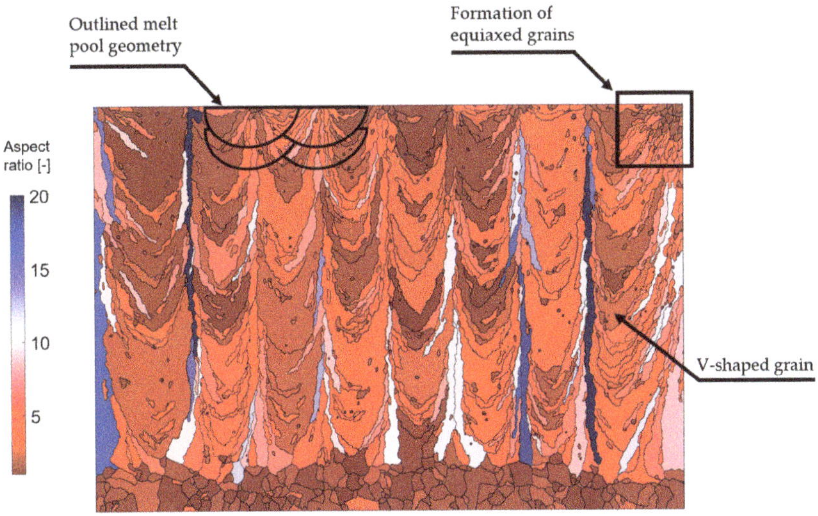


**Figure 8.** Aspect-ratio of the computed grain structure using $5 \times 10^{11}$ nucleation density. Some characteristics of the as-solidified grain texture are highlighted.

A zone of equiaxed grains is formed at the top-right of Figure 8. This is not credited to the choice of process parameters. It is rather because it is the last deposited hatch, see Figure 9c, where the cooling conditions and melt pool geometry changes in the FE model compared to the other hatches (hatches 2–8), see Figure 9b. For the last deposited hatch, the cooling conditions give more favorable conditions for equiaxed grains to grow. The first hatch also has different cooling conditions but equiaxed grains are not forming, see Figure 9a. It is also seen that both the first and the last hatch become wider than the intermediate hatches.

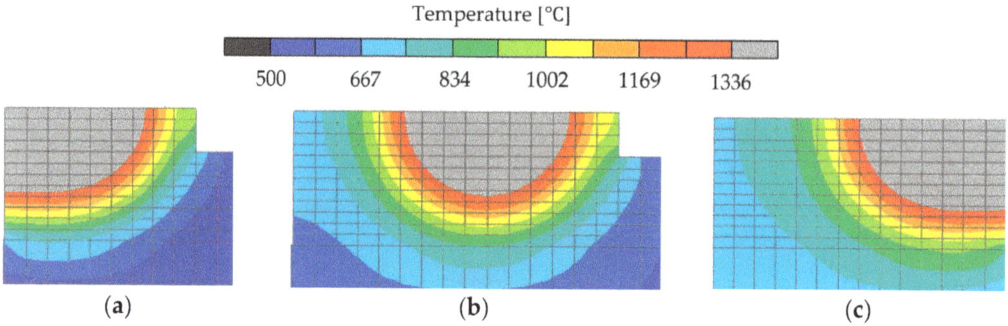


**Figure 9.** Three scenarios during deposition of hatches for the third layer giving different temperature gradients and rates on the studied BD-TD plane, (**a**) shows the first added hatch on the layer, (**b**) is the intermediate hatches, and (**c**) is the last added hatch on the layer. The gray zone shows the geometry of the melt pool.

### 3.2. Influence on Grain Texture by Nucleation Density

Figure 10 shows different types of simulated grain textures obtained when using different nucleation densities $n_{max}$. The top three layers of an already simulated texture were used as an initial substrate rather than having an equiaxed build plate as shown in Figure 8. This is because the simulation should represent a volume somewhere in the middle of the wall to be comparable with experiments from Malmelöv et al. [8]. It should be noted that the color fringe is difficult to compare between simulation and experiments since in the 2D model there is only one Euler angle giving the rotation of the crystal while in 3D three Euler angles give the rotation of the crystal. Instead, the general trend of the epitaxial grain growth is discussed rather than the trends of the crystallographic orientation. The calibration is referred to as case A where the scanning velocity was set to 1200 mm/s and the power was set to 100 W.

In Figure 10a, most of the grains grow epitaxially through all the 15 deposited layers when using a nucleation density of $1 \times 10^{11}$. Figure 10b shows a texture with most of the grains growing epitaxially through at least five of the deposited layers when using a nucleation density of $3 \times 10^{11}$. The texture for the nucleation density $5 \times 10^{11}$ in Figure 10c shows how the grain texture consists mostly of grains that do not grow epitaxially through five or more deposited layers. Even less epitaxial growth is observed for the nucleation density $10 \times 10^{11}$ in Figure 10d. These results show the significance of using the right nucleation density since the grain texture is largely influenced by this number, and thus this value must be calibrated to experimental observations. By visual comparison to the EBSD imaging from Malmelöv et al. [8], it is seen that $1 \times 10^{11}$ is too low for a nucleation density since no grain grew epitaxially through 15 layers in the experiment. The nucleation density $10 \times 10^{11}$ is too high since there were grains that grew epitaxially over more than five layers.

When comparing the trends of the crystallographic orientation in the simulated textures, it is seen for a nucleation density of $10 \times 10^{11}$ in Figure 10d that most larger grains have an orientation of 30 degrees or more. Grains that have a high misalignment of dendrite tip with the heat flow direction tend to become crowded out by the grains with more favorable orientations. These unfavorably oriented grains tend to be small and slender with a crystallographic orientation between 0 and 10 degrees. This shows that the competitiveness of grain growth is captured by the model. This trend is to some extent captured by the lower nucleation densities of $3 \times 10^{11}$ and $5 \times 10^{11}$, however, there are some larger grains with a high misalignment of the dendritic tip.

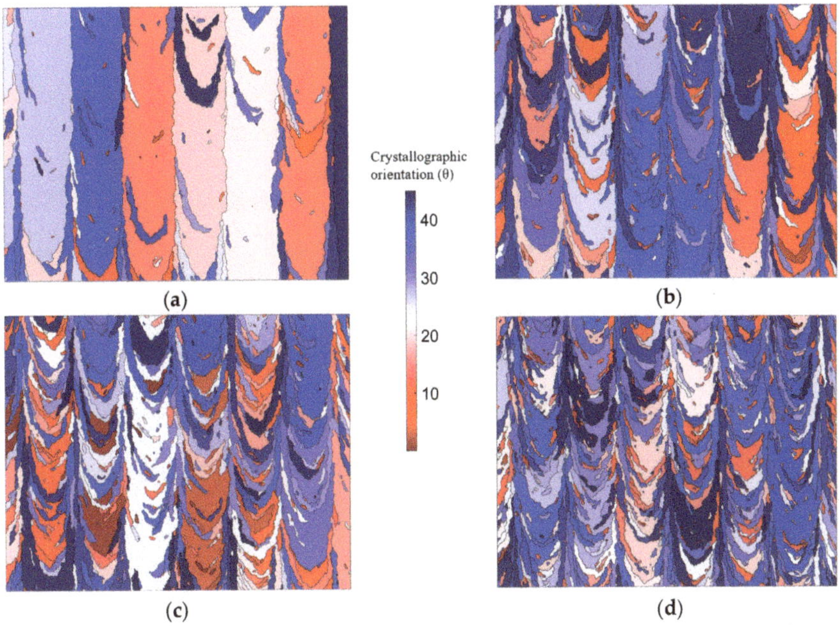


**Figure 10.** Simulated grain texture for different nucleation densities $n_{max}$ (**a**) $1 \times 10^{11}$, (**b**) $3 \times 10^{11}$, (**c**) $5 \times 10^{11}$, and (**d**) $10 \times 10^{11}$.

The aspect ratio was plotted versus the grain size in Figure 11a for three of the simulated grain textures, neglecting the texture for nucleation density $10 \times 10^{11}$ which contains too many small grains compared to the EBSD imaging from Malmelöv et al. [8]. The analysis was made using the MATLAB toolbox MTEX and only grains that contained 200 or more pixels were considered like the analysis in Malmelöv et al. [8]. It is seen that for all simulations most grains are smaller than 1500 $(\mu m)^2$ with an aspect ratio below 10 which agrees with the measurements.

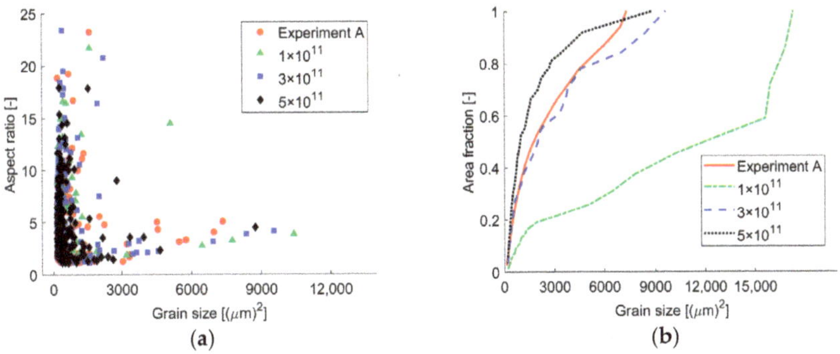


**Figure 11.** Comparison between simulated grain texture for different nucleation densities with experimental data adapted from Reference [8] where (**a**) shows the aspect ratio vs. grain size dot plot and (**b**) shows the grains' accumulated contribution to the total analyzed area.

The grain size distribution for grains smaller than 2000 $(\mu m)^2$ is illustrated in Figure 12. The grain texture with the nucleation density $1 \times 10^{11}$ predicts grain sizes that are too large compared to the measured values, see Figure 12a. For the nucleation density of $5 \times 10^{11}$, most of the grains are smaller than those of the experiments, see Figure 12c. The nucleation density $3 \times 10^{11}$ also predicts too small grains although to a much lesser extent than nucleation density $5 \times 10^{11}$. When comparing the grains' accumulated contribution to the total analyzed area in Figure 11b, it is seen that the nucleation density $3 \times 10^{11}$ agrees well with the experiment. This is also the reason why it was assumed that an appropriate value of the nucleation density is $3 \times 10^{11}$.

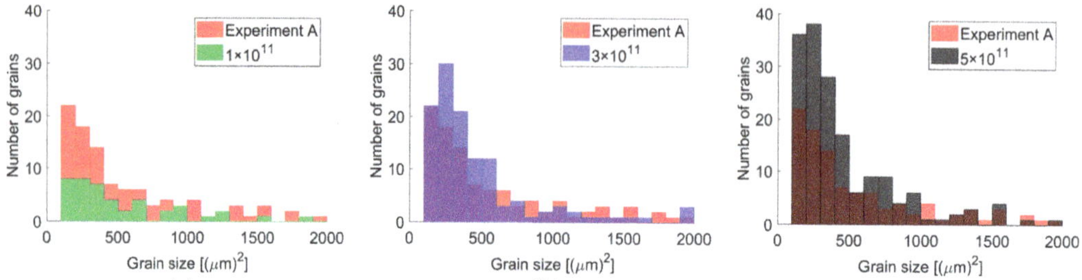


(a) (b) (c)

**Figure 12.** Grain size distribution for different nucleation densities (**a**) $1 \times 10^{11}$, (**b**) $3 \times 10^{11}$, and (**c**) $5 \times 10^{11}$ compared with the experimental data adapted from Reference [8].

### *3.3. Influence on Grain Texture by Changed Process Parameters*

As discussed in Section 3.1, the cooling conditions change for the first and last hatch compared to the intermediate hatches. For the second set of parameters only the intermediate zone with hatch numbers 2–8 is analyzed from the simulated grain textures where the cooling conditions are similar. This is referred to as case B with the scan velocity set to 1400 mm/s and the power set to 160 W. The upper 15 μm of the simulation domain is ignored since this zone contains small grains that would disappear if more layers were deposited. This gives a total analyzed area in agreement with the EBSD images in Malmelöv et al. [8]. Figure 13a shows the simulated grain texture for this set of parameters with more pronounced columnar grains compared to case A. Figure 13b shows the aspect ratio for each grain plotted versus its grain size for the simulated grain texture and experimental results, respectively. The general trend of the grain texture is that it contains larger grains compared to case A. Figure 13c shows the number of grains within an interval of grain sizes for case B. The discrepancy between the experimental results in the literature and the simulation can partially be explained by the local variation of the nucleation density in different parts of the build and therefore also the grain texture will vary.

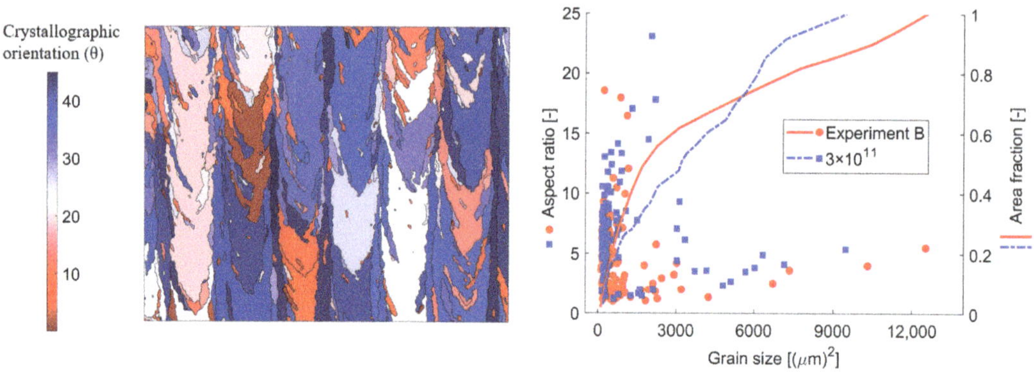


(a) (b)

**Figure 13.** *Cont.*

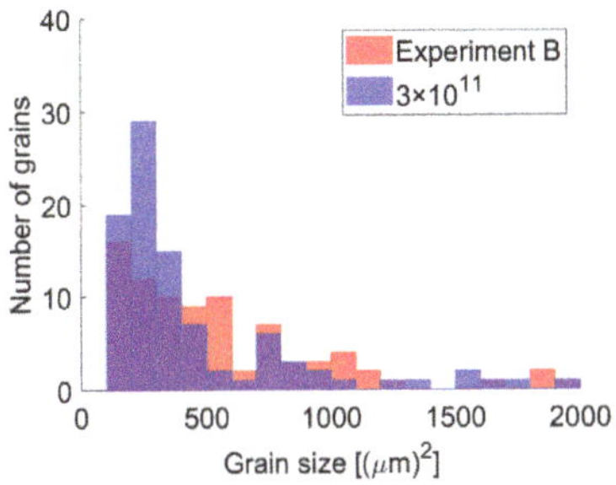


(c)

**Figure 13.** Simulated grain texture (**a**) with nucleation density $3 \times 10^{11}$ for case B, and comparison between simulation and the adapted experimental data from Reference [8] where (**b**) shows the aspect-ratio vs grain size dot plot and the grains' accumulated contribution to the total analyzed area and (**c**) shows the grain size distribution.

## 4. Conclusions

A combined finite element and cellular automata model (CA-FE) has been developed to predict the temperature history and grain texture of alloy 625 processed by PBF-LB. The general solidification behavior could be captured using the developed 2D CA-FE model. Epitaxial grain growth can be captured by the 2D CA-FE model, resulting in the columnar grains seen in additive manufactured products. The columnar grains form as coarse grains in the overlap between two adjacent melt pool boundaries, whereas slender elongated grains form in the center of the melt pool. Equiaxed grain growth is also captured by the 2D CA-FE model. The equiaxed grains form when grains nucleated through bulk nucleation suppress the growth of the columnar growth front. The 2D CA-FE model can capture the competitiveness of grain growth where more favorably oriented grains outcompete less favorably oriented grains. The 2D CA-FE model can capture changes in grain texture depending on the process parameters. A higher scan velocity and power resulted in a grain texture with larger grains.

It is seen that the 2D CA-FE model simulates the grain texture which agrees well with the experimental findings in the literature. One essential parameter that has a high influence on the predicted grain texture is the nucleation density. It was shown that comparison to EBSD images can be used to determine suitable nucleation parameters. Although grain growth and nucleation are a 3D phenomenon, a 3D CA-FE model increases the computational time drastically. These computational times are too long for the model to be useful by engineers during process development. It has been shown that the 2D CA-FE model can be used to evaluate the as-printed grain texture depending on the chosen process parameters for PBF-LB such as scan velocity and laser power. The FE simulation runs in about 20 h while the CA model takes about 8 h. The CA computational time could most likely be considerably reduced by parallelization and implementation in a compiled programming language such as Fortran.

**Author Contributions:** Conceptualization, C.A. and A.L; methodology, C.A. and A.L.; software, C.A.; validation, C.A.; formal analysis, C.A.; investigation, C.A.; resources, A.L.; data curation, C.A.; writing—original draft preparation, C.A. and A.L.; writing—review and editing, C.A. and A.L.; visualization, C.A.; supervision, A.L.; project administration, A.L.; funding acquisition, A.L. All authors have read and agreed to the published version of the manuscript.

**Funding:** This work is partially carried out within the FINAST project (Research and Innovation in Norrbotten for Advanced Green Steel Production and Manufacturing) which is funded by the EU Just Transition Fund and the Swedish Agency for Economic and Regional Growth under grant number 20358499. The Creaternity network at LTU is also acknowledged for their partial funding.

**Data Availability Statement:** The data presented in this study are available on request from the corresponding author.

**Conflicts of Interest:** The authors declare no conflict of interest.

## References


1. Ford, S.; Despeisse, M. Additive Manufacturing and Sustainability: An Exploratory Study of the Advantages and Challenges. *J. Clean. Prod.* **2016**, *137*, 1573–1587. [CrossRef]
2. Petrovic, V.; Gonzalez, J.V.H.; Ferrando, O.J.; Delgado Gordillo, J.; Puchades, J.R.B.; Griñan, L.P. Additive Layered Manufacturing: Sectors of Industrial Application Shown through Case Studies. *Int. J. Prod. Res.* **2011**, *49*, 1061–1079. [CrossRef]
3. Volpato, G.M.; Tetzlaff, U.; Fredel, M.C. A Comprehensive Literature Review on Laser Powder Bed Fusion of Inconel Superalloys. *Addit. Manuf.* **2022**, *55*, 102871. [CrossRef]
4. Dinda, G.P.; Dasgupta, A.K.; Mazumder, J. Laser Aided Direct Metal Deposition of Inconel 625 Superalloy: Microstructural Evolution and Thermal Stability. *Mater. Sci. Eng. A* **2009**, *509*, 98–104. [CrossRef]
5. Shankar, V.; Bhanu Sankara Rao, K.; Mannan, S.L. Microstructure and Mechanical Properties of Inconel 625 Superalloy. *J. Nucl. Mater.* **2001**, *288*, 222–232. [CrossRef]
6. Yan, X.; Gao, S.; Chang, C.; Huang, J.; Khanlari, K.; Dong, D.; Ma, W.; Fenineche, N.; Liao, H.; Liu, M. Effect of Building Directions on the Surface Roughness, Microstructure, and Tribological Properties of Selective Laser Melted Inconel 625. *J. Mater. Process. Technol.* **2021**, *288*, 116878. [CrossRef]
7. Li, C.; White, R.; Fang, X.Y.; Weaver, M.; Guo, Y.B. Microstructure Evolution Characteristics of Inconel 625 Alloy from Selective Laser Melting to Heat Treatment. *Mater. Sci. Eng. A* **2017**, *705*, 20–31. [CrossRef]
8. Malmelöv, A.; Hassila, C.J.; Fisk, M.; Wiklund, U.; Lundbäck, A. Numerical Modeling and Synchrotron Diffraction Measurements of Residual Stresses in Laser Powder Bed Fusion Manufactured Alloy 625. *Mater. Des.* **2022**, *216*, 110548. [CrossRef]
9. Kou, S. *Welding Metallurgy*; Wiley: Hoboken, NJ, USA, 2002.
10. Andreau, O.; Koutiri, I.; Peyre, P.; Penot, J.D.; Saintier, N.; Pessard, E.; De Terris, T.; Dupuy, C.; Baudin, T. Texture Control of 316L Parts by Modulation of the Melt Pool Morphology in Selective Laser Melting. *J. Mater. Process. Technol.* **2019**, *264*, 21–31. [CrossRef]
11. Flood, S.C.; Hunt, J.D. Columnar and Equiaxed Growth: II. Equiaxed Growth Ahead of a Columnar Front. *J. Cryst. Growth* **1987**, *82*, 552–560. [CrossRef]
12. Rappaz, M.; Gandin, C.A. Probabilistic Modelling of Microstructure Formation in Solidification Processes. *Acta Metall. Mater.* **1993**, *41*, 345–360. [CrossRef]
13. Körner, C.; Markl, M.; Koepf, J.A. Modeling and Simulation of Microstructure Evolution for Additive Manufacturing of Metals: A Critical Review. *Metall. Mater. Trans. A* **2020**, *51*, 4970–4983. [CrossRef]
14. Kurz, W.; Rappaz, M.; Trivedi, R. Progress in Modelling Solidification Microstructures in Metals and Alloys. Part II: Dendrites from 2001 to 2018. *Int. Mater. Rev.* **2020**, *66*, 30–76. [CrossRef]
15. Gong, X.; Chou, K. Phase-Field Modeling of Microstructure Evolution in Electron Beam Additive Manufacturing. *JOM* **2015**, *67*, 1176–1182. [CrossRef]
16. Rodgers, T.M.; Madison, J.D.; Tikare, V. Simulation of Metal Additive Manufacturing Microstructures Using Kinetic Monte Carlo. *Comput. Mater. Sci.* **2017**, *135*, 78–89. [CrossRef]
17. Gandin, C.A.; Rappaz, M. A 3D Cellular Automaton Algorithm for the Prediction of Dendritic Grain Growth. *Acta Mater.* **1997**, *45*, 2187–2195. [CrossRef]
18. Gandin, C.A.; Rappaz, M. A Coupled Finite Element-Cellular Automaton Model for the Prediction of Dendritic Grain Structures in Solidification Processes. *Acta Metall. Mater.* **1994**, *42*, 2233–2246. [CrossRef]
19. Dezfoli, A.R.A.; Lo, Y.L.; Raza, M.M. Prediction of Epitaxial Grain Growth in Single-Track Laser Melting of IN718 Using Integrated Finite Element and Cellular Automaton Approach. *Materials* **2021**, *14*, 5202. [CrossRef]
20. Rai, A.; Markl, M.; Körner, C. A Coupled Cellular Automaton–Lattice Boltzmann Model for Grain Structure Simulation during Additive Manufacturing. *Comput. Mater. Sci.* **2016**, *124*, 37–48. [CrossRef]
21. Koepf, J.A.; Soldner, D.; Ramsperger, M.; Mergheim, J.; Markl, M.; Körner, C. Numerical Microstructure Prediction by a Coupled Finite Element Cellular Automaton Model for Selective Electron Beam Melting. *Comput. Mater. Sci.* **2019**, *162*, 148–155. [CrossRef]
22. Teferra, K.; Rowenhorst, D.J. Optimizing the Cellular Automata Finite Element Model for Additive Manufacturing to Simulate Large Microstructures. *Acta Mater.* **2021**, *213*, 116930. [CrossRef]
23. Isoltermics, Nicrofer6020 HMo—Alloy 625. Available online: https://www.isolthermics.com.au/metals/pdfs/common-tech/Alloy%20625%206020hMo.pdf (accessed on 18 August 2023).
24. Special Metals, Inconel Alloy 625. Available online: https://www.specialmetals.com/documents/technical-bulletins/inconel/inconel-alloy-625.pdf (accessed on 18 August 2023).
25. Denlinger, E.R.; Michaleris, P. Effect of Stress Relaxation on Distortion in Additive Manufacturing Process Modeling. *Addit. Manuf.* **2016**, *12*, 51–59. [CrossRef]
26. Wei, L.C.; Ehrlich, L.E.; Powell-Palm, M.J.; Montgomery, C.; Beuth, J.; Malen, J.A. Thermal Conductivity of Metal Powders for Powder Bed Additive Manufacturing. *Addit. Manuf.* **2018**, *21*, 201–208. [CrossRef]

27. Tinoco, J.; Fredriksson, H. Solidification of a Modified Inconel 625 Alloy under Different Cooling Rates. *High Temp. Mater. Process.* **2004**, *23*, 13–24. [CrossRef]
28. Goldak, J.; Chakravarti, A.; Bibby, M. A New Finite Element Model for Welding Heat Sources. *Metall. Trans. B* **1984**, *15*, 299–305. [CrossRef]
29. Lindgren, L.-E. *Computational Welding Mechanics: Thermomechanical and Microstructural Simulations*; Woodhead Publishing Ltd.: Cambridge, UK, 2007; ISBN 9781845692216.
30. Mohebbi, M.S.; Ploshikhin, V. Implementation of Nucleation in Cellular Automaton Simulation of Microstructural Evolution during Additive Manufacturing of Al Alloys. *Addit. Manuf.* **2020**, *36*, 101726. [CrossRef]
31. Lipton, J.; Glicksman, M.E.; Kurz, W. Dendritic Growth into Undercooled Alloy Metals. *Mater. Sci. Eng.* **1984**, *65*, 57–63. [CrossRef]
32. Kurz, W.; Giovanola, B.; Trivedi, R. Theory of Microstructural Development during Rapid Solidification. *Acta Metall.* **1986**, *34*, 823–830. [CrossRef]
33. Zinoviev, A.; Zinovieva, O.; Ploshikhin, V.; Romanova, V.; Balokhonov, R. Evolution of Grain Structure during Laser Additive Manufacturing. Simulation by a Cellular Automata Method. *Mater. Des.* **2016**, *106*, 321–329. [CrossRef]
34. Tsai, D.C.; Hwang, W.S. A Three Dimensional Cellular Automaton Model for the Prediction of Solidification Morphologies of Brass Alloy by Horizontal Continuous Casting and Its Experimental Verification. *Mater. Trans.* **2011**, *52*, 787–794. [CrossRef]
35. Wang, W.; Lee, P.D.; McLean, M. A Model of Solidification Microstructures in Nickel-Based Superalloys: Predicting Primary Dendrite Spacing Selection. *Acta Mater.* **2003**, *51*, 2971–2987. [CrossRef]
36. Yang, M.; Wang, L.; Yan, W. Phase-Field Modeling of Grain Evolutions in Additive Manufacturing from Nucleation, Growth, to Coarsening. *NPJ Comput. Mater.* **2021**, *7*, 56. [CrossRef]
37. Wang, L.Y.; Wang, Y.C.; Zhou, Z.J.; Wan, H.Y.; Li, C.P.; Chen, G.F.; Zhang, G.P. Small Punch Creep Performance of Heterogeneous Microstructure Dominated Inconel 718 Fabricated by Selective Laser Melting. *Mater. Des.* **2020**, *195*, 109042. [CrossRef]

metals

MDPI

Article

# Extending Density Phase-Field Simulations to Dynamic Regimes

**David Jacobson [1], Reza Darvishi Kamachali [2] and Gregory Bruce Thompson [1,*]**

[1] Department of Metallurgical and Materials Engineering, University of Alabama, Tuscaloosa, AL 35487, USA; dwjacobson@crimson.ua.edu
[2] Federal Institute for Materials Research and Testing, 12205 Berlin, Germany; reza.kamachali@bam.de
* Correspondence: gthompson@eng.ua.edu

**Abstract:** Density-based phase-field (DPF) methods have emerged as a technique for simulating grain boundary thermodynamics and kinetics. Compared to the classical phase-field, DPF gives a more physical description of the grain boundary structure and chemistry, bridging CALPHAD databases and atomistic simulations, with broad applications to grain boundary and segregation engineering. Notwithstanding their notable progress, further advancements are still warranted in DPF methods. Chief among these are the requirements to resolve its performance constraints associated with solving fourth-order partial differential equations (PDEs) and to enable the DPF methods for simulating moving grain boundaries. Presented in this work is a means by which the aforementioned problems are addressed by expressing the density field of a DPF simulation in terms of a traditional order parameter field. A generic DPF free energy functional is derived and used to carry out a series of equilibrium and dynamic simulations of grain boundaries in order to generate trends such as grain boundary width vs. gradient energy coefficient, grain boundary velocity vs. applied driving force, and spherical grain radius vs. time. These trends are compared with analytical solutions and the behavior of physical grain boundaries in order to ascertain the validity of the coupled DPF model. All tested quantities were found to agree with established theories of grain boundary behavior. In addition, the resulting simulations allow for DPF simulations to be carried out by existing phase-field solvers.

**Keywords:** density phase field; grain boundaries; numerical methods

## 1. Introduction

Grain boundaries play an out-sized role in determining material properties in metals such as strength, electrical and thermal conduction, etc. A significant amount of research has been conducted to determine how particular grain boundary arrangements can be achieved through different synthesis processes and conditions in order to obtain desired properties in polycrystalline materials. A more difficult problem has been determining how to predict and control the evolution of grain boundary networks over time. The ability to accurately model the evolution of grain boundaries is necessary for being able to predict how metallic material properties evolve in response to heat, stress, magnetic fields, and other phenomena [1]. Such modelling has been facilitated by the advent of high-performance computer systems which now alloww for in silico experiments to extend from the quantum scale to system/macro level simulations.

Many of today's open problems in materials science have to do with limitations to computational modelling associated with length and timescales. The larger the system modelled, the more underlying physics is often neglected or simplified in order to make the problem computationally feasible. The problem of grain boundary network evolution falls into such a category. Classical phase-field methods have emerged on the mesoscale describing microstructure evolution [2]. Significant progress has been made in phase-field modeling of grain boundary motion in polycrystalline materials, especially with the development of the multi-phase-field methods to explore grain boundary junctions, vertexes and their dynamics in various setups [3–5]. Yet, grain boundaries exhibit an incredible level of detail and variation in behavior at the atomic scale (low-angle grain boundary arrangements, special character boundaries, random high-angle boundaries, twist vs. tilt configurations, etc.). Atomistic simulations can capture such details of many if not all of these grain boundary behaviors but are computationally too expensive for studying the influence of such phenomena on the time scales of grain boundary network coarsening. The classical phase-field simulations sacrifice much of the aforementioned details of grain boundary physics in order to operate at lengths and timescales at which grain boundary coarsening can be observed, relying on phenomenological descriptions of grain boundaries as opposed to physical models. An overview of the application of phase-field methods to grain growth problems is given in the following references [6–10]. A broader overview of traditional phase-field theory and capabilities can be found in the following references [3,4,11,12].

To overcome some of these challenges, density phase-field (DPF) methods have arisen as an attractive alternative for describing grain boundaries in a more physical manner than traditional phase-field methods based on logistic-type order parameters ($\phi$) [13–18]. The idea behind DPF simulations is that the free energy of a grain boundary is more directly related to the atomic density at the grain boundary as opposed to traditional order parameters. The reduced density and disordered atomic environments at grain boundaries result in significant bond strains that are the ultimate source of the excess grain boundary energy. The atomic density order parameter enables DPF simulations to be integrated with CALPHAD databases while strongly linked with atomistic simulations in a physically-sound manner. The resulting density-based free energy functional is able to capture the temperature and composition dependence of grain boundary energetics, giving an accurate description of grain boundary phase behavior that may not present in traditional phase-field simulations [19,20]. The remainder of this introduction provides a brief overview of the types of DPF simulations available, as well as a discussion on the current technical challenges associated with carrying out DPF simulations.

### 1.1. Overview of Density Phase-Field Theory

DPF solvers work similarly to their classical counterparts in that they minimize a free energy functional of the form given by Equation (1) by solving a model A type equation for unconserved dynamics (Equation (2)).

$$F = \int_v F_v(\rho, \nabla\rho, \dots)\, dV \tag{1}$$

$$\frac{\partial \rho}{\partial t} = -m_\rho \mu_\rho \tag{2}$$

with the potential function $\mu_\rho = \frac{\delta F_v}{\delta \rho}$. Typically, a normalized form of atomic density is used such that $\rho = 1$ at equilibrium associated with the parent bulk phase. One can calculate the real atomic density simply by multiplying the reference bulk atomic density being studied ($\rho^0_{atom}$) by the local value of the normalized density ($\rho_{atomic} = \rho^0_{atomic}\rho$). To avoid confusion between $\rho$ and $\rho_{atom}$, the reference atomic density is expressed as the inverse of the molar volume ($\rho_{atom} = \hat{V}_0^{-1}$). The form of the volumetric free energy function $F_v$ varies between models. In its most basic and general form, it mirrors classical phase field free energy functionals in that it is composed of two terms: a bulk term to describe the free energy of the system as a function of local density plus a gradient energy series to take into account the energy associated with spatial variations of $\rho$. The general form of the volumetric free energy functional is given below.

$$F_v = f_{bulk}(\rho) + \sum_i \kappa_i |\nabla^i \rho|^2 \tag{3}$$

Kamachali derived the volumetric free energies as a deviation from the bulk free energy curve of a solid solution [13]. For a regular solution, the free energy is presented in Equation (4)

$$F_v = X_A\left(E_A^B\rho^2 + (K_A^B + PV_A - TS_A^B)\rho + \kappa_{A,1}|\nabla\rho|^2 + \kappa_{A,2}|\nabla^2\rho|^2\right) + X_B\left(E_B^B\rho^2 + (K_B^B + PV_B - TS_B^B)\rho + \kappa_{B,1}|\nabla\rho|^2 + \kappa_{B,2}|\nabla^2\rho|^2\right) + \rho^2\Omega X_A X_B - T\Delta S_{mix}^B + \kappa_X|\nabla X_B|^2 \quad (4)$$

Equation (4) resembles the standard CALPHAD formulation, upgraded with the density-dependent terms and the gradient energy terms. Compared to the classical phase-field models [3], a major advantage of DPF approach is that it allows for the natural development of grain boundary free energy functional, integrated with the CALPHAD framework. This allowed for the successful prediction of spinodal decomposition occurring at the grain boundaries outside of the bulk miscibility gap in the iron alloys [14,19] as well as in the platinum-gold system [15]. Jacobson et al. took a different approach by using atomistic theory to derive a free energy functional based on interatomic potentials [21]. The resulting class of simulations has been termed the Molecular Phase-Field method (MoPF); the general and Morse forms of which are given by Equations (5) and (6).

$$F_v = \frac{\rho}{\hat{V}_0}\sum_i \frac{n_i}{2}U_{bond}(\rho) + \sum_i \kappa_i|\nabla^i\rho|^2 \quad (5)$$

$$F_v = \frac{\rho}{\hat{V}_0}\sum_i \frac{n_i}{2}\epsilon\left[e^{-2\alpha(r_i^*\rho^{-\frac{1}{3}}-r_0)} - 2e^{-\alpha(r_i^*\rho^{-\frac{1}{3}}-r_0)}\right] + \sum_i \kappa_i|\nabla^i\rho|^2 \quad (6)$$

The primary advantage of the MoPF model is that the interatomic potential parameters used as inputs to the model naturally convey material-specific characteristics to the model. For example, in the absence of gradients, the MoPF method correctly predicts the bulk modulus of the material. For grain boundaries, the grain boundary free energy is a natural consequence of the model rather than an objective value that the model must be calibrated for in order to reproduce. However, the atomistic simulations can be computationally expensive.

There are two key differences between classical phase-field and DPF methods. In a classical phase-field simulation, the order parameter $\phi$ represents crystallographic misorientation, but the association between the misorientation and the order parameter is rather arbitrary. The atomic density field on the other hand is physically linked with the substructure of the grain boundary. This constraint is taken into account by modifying the model A equation. Typically, the model A equation relies on a potential function $\mu_\rho$ that follows a non-conserved variational form given below.

$$\mu_\rho = \frac{\delta F}{\delta\rho} = \frac{\partial F_v}{\partial\rho} - \nabla\frac{\partial F_v}{\partial\nabla\rho} \quad (7)$$

Considering the mass conservation, the evolution of the density field shall be taken as below, taking $\Delta\mu_\rho$ instead of $\mu_\rho$.

$$\frac{\partial\rho}{\partial t} = -m_\rho\Delta\mu_\rho \quad (8)$$

Here, the potential difference indicates the relation between the change in density and mass transfer into or out of the boundary. Assumed in the density-based model is that the source/sink of these atoms corresponds to a reservoir consisting of a perfect crystal at constant density ($\rho = 1$) where the density potential $\mu_\rho = \mu_\rho^0$. In such a reservoir, the gradient contribution to the density potential is zero. If we assume that the energy change due to a change in reservoir volume is negligible (a change in reservoir volume is necessary to keep the density constant), the resulting form of $\mu_\rho^0$ is simply the volumetric free energy of an unstrained perfect crystal as is shown in Equation (9).

$$\Delta\mu_\rho = \frac{\delta F}{\delta\rho}(\rho, \nabla\rho, ...) - F_v(\rho = 1, \nabla\rho = 0, ...) \quad (9)$$

In reality, density changes in a grain boundary typically occur through vacancy emission and absorption [22]. The reservoir concept defined above is meant to provide a simpler and more computationally efficient means of modelling grain boundary density dynamics than attempting to model vacancy migration, generation, and elimination at the mesoscale.

Another key difference between the DPF methods and classical phase-field simulations deals with what we term the center boundary condition. Grain boundaries are non-equilibrium defects that are inherently unstable. If not for the existence of a large activation barrier associated with grain rotation, grains would simply reorient themselves such that all grain boundaries were eliminated. The aforementioned activation barrier is a direct result of the crystallographic misorientation between the two grains composing the boundary. Unfortunately, this activation barrier is not captured in DPF models which necessitates a "center boundary condition", i.e., the density at the grain boundary center must be specified and held constant throughout the simulation. If this was not the case, grain boundaries in DPF simulations would simply dissipate until the normalized density everywhere equaled one (the equilibrium value). The density at the center boundary condition is the lowest density in the entire simulation and is referred to as $\rho_{min}$. The choice of $\rho_{min}$ can be based on a variety of criteria but is most often meant to relate with the misorientation angle [13] and to match the minimum calculated density of a grain boundary generated using molecular dynamics.

### 1.2. *Issues with the Density Phase-Field Method*

#### 1.2.1. Theoretical Issues

The DPF model was derived assuming that only attractive interatomic forces were needed. By doing so, this greatly simplifies the complexity of calculating $\mu_\rho$, but at the expense of thermodynamic consistency. We illustrate this problem using the single component free energy functional and its respective density potential difference given below.

$$F_v = \rho(\rho\hat{E}_A + \hat{K}_A + P\hat{V}_A - T\hat{S}_A) + \kappa_\rho|\nabla\rho|^2 \quad (10)$$

$$\Delta\mu_\rho = (2\rho - 1)\hat{E}_A - \kappa_\rho\nabla^2\rho \quad (11)$$

In the absence of any gradients, $\Delta\mu_\rho(\rho = 1)$ should equal zero because $\rho = 1$ is defined as the equilibrium condition. Due to the simple linear form of the potential energy, this condition is not naturally met and is imposed by the condition of having $\rho \leq 1$ throughout the system. While being a robust solution for studying static grain boundaries, we have found that such enforcement schemes introduce non-physical behaviors into dynamic simulations. This motivated our desire to improve the density-based free energy functional such that the equilibrium condition is met in a dynamic state.

#### 1.2.2. Computational Challenges

There is a computational challenge associated with the center boundary condition. Although the center boundary condition is necessary for grain boundaries to exist in DPF simulations, it makes simulating the motion of grain boundaries difficult. The question arises as to how can the center boundary condition be moved such that the density field evolves in a natural manner? Although distinct from DPF, the authors point to the work by Phillippe et al. to further illustrate the numerical difficulties that arise from solving the evolution equations of grain boundaries using non-trivial free energy functionals [23].

The gradient energy terms in the DPF free energy functional are another source of difficulty in carrying out DPF simulations. Classical phase-field models are limited to a single first-order term in gradient energy. However, higher-order gradient terms are known to significantly improve the accuracy of phase-field models as well as the DPF model. Additionally, the resulting higher-order partial differential equations (PDEs) are much stiffer and more computationally intensive to solve than lower-order PDEs. The DPF simulations are shown to give a smooth density profile with a minimum of first- and second-order gradient terms. This fact is demonstrated in Figure 1, one computed with only the first-order gradient term (plot A) and another with both first- and second-order terms (plot B). As a result, the model A equation for DPF simulations is a fourth-order PDE that is significantly more difficult and time-consuming to solve numerically than it is for classical phase-field simulations.

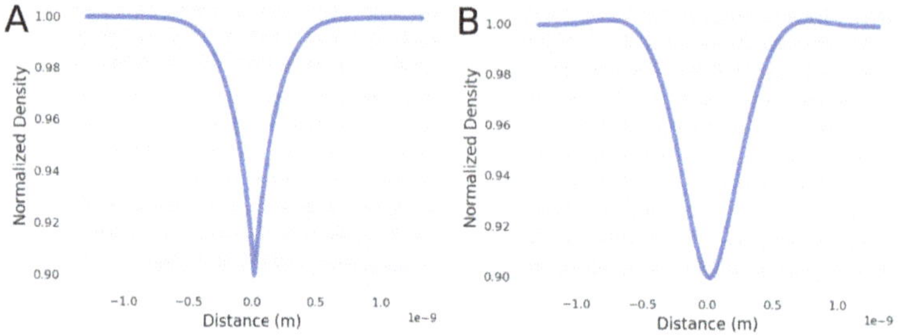


**Figure 1.** Plot (**A**): the density field of a non-coupled DPF simulation where only the first term of the gradient energy series is included. Plot (**B**): the density field of a non-coupled DPF simulation where the first two terms of the gradient energy series are included. Notice that the inclusion of the higher-order gradient energy term eliminates the sharp point at the grain boundary center. The inclusion of the of the higher-order gradient energy terms also results in the larger than bulk density regions at the grain boundary periphery.

In this study, we seek to address the issues outlined above and answer the following questions:

- What are the thermodynamic criteria that density phase field free energy functionals should meet?
- How can the DPF methods be made dynamic?
- How can the performance constraints associated with solving a fourth-order PDE be overcome?

## 2. Theory

### *2.1. General Criteria for Density Free Energy Functionals*

This section is dedicated to answering question one from the introduction. Generally, a thermodynamically consistent density free energy functional needs to, at a minimum, meet the following criteria:

1. $F_v(\rho = 1, \nabla\rho = 0) = F_v^0$
2. $\Delta\mu_\rho(\rho = 1, \nabla\rho = 0) = 0$
3. $\frac{\partial\mu_\rho}{\partial\rho}(\rho = 1, \nabla\rho = 0) > 0$
4. $F_v(\rho = 0, \nabla\rho = 0) = 0$
5. $\frac{\partial F_v}{\partial\rho}(\rho = 0, \nabla\rho = 0) = 0$

The first of these criteria is the continuity condition stating that the free energy should align with the bulk value $F_v^0$, whenever the excess density and its gradients vanish. The second criteria ensures that the bulk state of $(\rho = 1, \nabla\rho = 0)$ is a true equilibrium state. The third criteria ensures that the bulk equilibrium is stable. Criteria 4 and 5 are referred to as the sparse particle conditions. They ensure that atoms that are far apart have a negligible influence on one another. It should be noted that these criteria are useful for making the volumetric free energy more physically accurate across its entire domain, but have limited effect on the region of the free energy curve $(\rho \approx 1)$ relevant to grain boundary simulations.

The original density-based formulation does not meet the second and third criteria because repulsive interatomic contributions were not included. As noted above, such interactions can be explicitly modelled using the approach taken by Jacobson et al. by constructing the volumetric free energy functional from interatomic potentials. Here, we take a simpler approach by deriving a polynomial from of the bulk component of the volumetric free energy that satisfies criteria 1–5. For a grain boundary at $T$ = 0 K, the kinetic and entropic portions of the free energy functional are zero. We further assume that the pressure volume contribution is small enough to be neglected. Thus, the volumetric free energy can be expressed purely in terms of the molar potential energy $\hat{E}_A$ ($F_{v,bulk} = \frac{\rho}{\hat{V}_0}\hat{E}_A(\rho)$.) Recognizing that $\hat{E}_A(\rho = 1)$ is the cohesive energy ($E_{co}$), one can write

$$F_{v,bulk} = \rho E_{co} f(\rho) \tag{12}$$

Equation (12) meets criteria 1–3 when $f(\rho)$ gives $f(\rho = 1) = 1$, $\frac{\partial f}{\partial \rho}|_{\rho=1} = 0$ and $\frac{\partial^2 f}{\partial \rho^2}|_{\rho=1} < 0$ as $E_{co} < 0$. Infinitely many different polynomial forms can be constructed to meet such criteria. Under the premise that simpler is better, we use the following form of $f(\rho)$.

$$f(\rho) = \rho^n(a\rho^2 + b\rho + c) \tag{13}$$

Parameters $n$ and $a$ are free variables that can be used to change the potential well shape which corresponds to a change in the bulk properties of the material being modelled. The characteristic shape of Equation (13) is shown in Figure 2 where the curves illustrated demonstrate the influence of parameters $a$ and $n$ on the well shape. Once parameters $n$ and $a$ are chosen, $b$ and $c$ can be determined through the following equations.

$$b = -(n + 2a) \tag{14}$$

$$c = 1 + a + n \tag{15}$$

A more thorough derivation of the above two expressions is provided in the Appendix A.

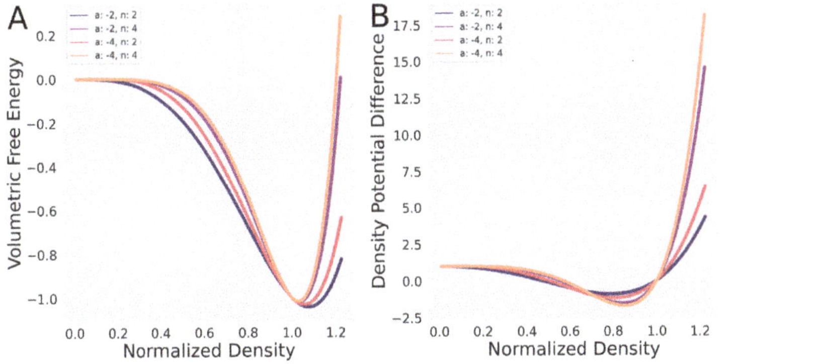


**Figure 2.** Plot (**A**): the bulk volumetric free energy function from Equations (12) and (13). Plot (**B**): the bulk component of the density potential difference corresponding to to Equations (12) and (13). Values of the cohesive energy and bulk modulus are −1 and 1, respectively ($E_{co} = -1$, $\hat{V}_0^{-}1 = 1$). Parameters $a$ and $n$ are varied to show their influence on the free energy curve well shape. Notice that for every curve the density potential difference at a density of one equals zero. This is by design to ensure that the bulk state ($\rho = 1$) is a stable equilibrium.

## 2.2. Linking Order Parameters with Density

Questions two and three can be solved by coupling the density phase-field method with traditional order parameters. We reintroduce the order parameter $\phi$ into the simulation as an independent field variable and make the density dependent on $\phi$ according to Equation (16)

$$\rho = 1 - 4(1 - \rho_{min})\phi(1 - \phi) \tag{16}$$

Doing so resolves the issues of the center boundary condition because the density field is naturally at a minimum where $\phi = 0.5$. Now the temporal evolution of the density variable can be achieved by solving the model A equation with respect to $\phi$ and recalculating $\rho$ using Equation (16) after every time step. The evolution equation is derived below.

$$\frac{\partial \phi}{\partial t} = m_\phi \left( \frac{\delta F}{\delta \phi} - \mu_\phi^0 \right) \tag{17}$$

$$\frac{\delta F}{\delta \phi} = \frac{\partial F_v}{\partial \phi} - \nabla \frac{\partial F_v}{\partial \nabla \phi} \tag{18}$$

The reference component of Equation (17) ($\mu_\phi^0$) can be calculated simply by appending the derivative of $\rho$ with respect to $\phi$ to the equivalent reference component in Equation (9).

$$\mu_\phi^0 = F_v(\rho = 1, \nabla \rho = 0) \frac{\partial \rho}{\partial \phi} \tag{19}$$

The first term of Equation (18) can be solved for through use of the chain rule.

$$\frac{\partial F_v}{\partial \phi} = \frac{\partial F_v}{\partial \rho} \frac{\partial \rho}{\partial \phi} \tag{20}$$

A direct connection between the mobilities ($m_\phi$ and $m_\rho$) is not possible because the use of the center boundary condition when carrying out density dynamics prevents the expression $\frac{\partial \rho}{\partial t} = \frac{\partial \rho}{\partial \phi} \frac{\partial \phi}{\partial t}$ from being valid over the entire domain. The second term of Equation (18) can be solved for explicitly by substituting Equation (16) into the gradient free energy series. Doing so increases the complexity of the free energy functional though without alleviating the large performance constraints associated with solving a fourth-order PDE. Instead, we motivate an approximation that simplifies the free energy functional in addition to making the evolution equation second order.

The gradient energy term is meant to take into account the excess energy resulting from non-equilibrium environments associated with the spatial transition of a field variable. For the case of DPF simulations, the gradient energy is meant to take into account the disordered bonding environment found at grain boundaries. Atomistic simulations indicate that the greatest degree of disorder is found at the center of the grain boundary as opposed to at its periphery. Thus, we can surmise that the gradient energy series should predict a maximum free energy at the grain boundary center. This is why gradient energy terms with orders in excess of one are required for non-coupled DPF simulations. At the grain boundary center, the density gradient must be zero and by extension the first order term of the gradient energy series is zero. Even when a second order term is included, there are still non-physical artifacts in the resulting density field. One will notice in Figure 1 that the sharp grain boundary center is eliminated by the inclusion of the second order term, but that the density is now overestimated in the periphery of the grain boundary where the density values exceed 1. It is possible that the addition of more higher-order terms would eventually yield a satisfactory density profile, but determining the coefficients for said terms would be non-trivial and the resulting PDEs would be stiff and require time step sizes that are too small to be considered practical. As a result, we make the following simplifying assumption.

$$\sum_i \kappa_i |\nabla^i \rho|^2 \approx \kappa_\phi |\nabla \phi|^2 \tag{21}$$

The motivation for this assumption is two-fold. From a computational stand point, it reduces the evolution equation to second order and simplifies the free energy functional.

From a theoretical stand point, we simply note that the density profiles obtained using the approximation given by Equation (21) more closely match density profiles obtained from atomistic simulations than density profiles obtained using the the density gradient energy sum. Examples of density profiles obtained from atomistic simulations are shown in the second figure of the work by Jacobson et al. [21]. The fully coupled form of the volumetric free energy functional that will be used throughout the remainder of this work is given by Equation (22)

$$F_v = \frac{\rho^{n+1}}{\hat{V}_0} E_{co}(a\rho^2 + b\rho + c) + \kappa_\phi |\nabla\phi|^2. \tag{22}$$

The order parameter and density profiles resulting from a free energy expression, such as Equation (22), are given in Figure 3.

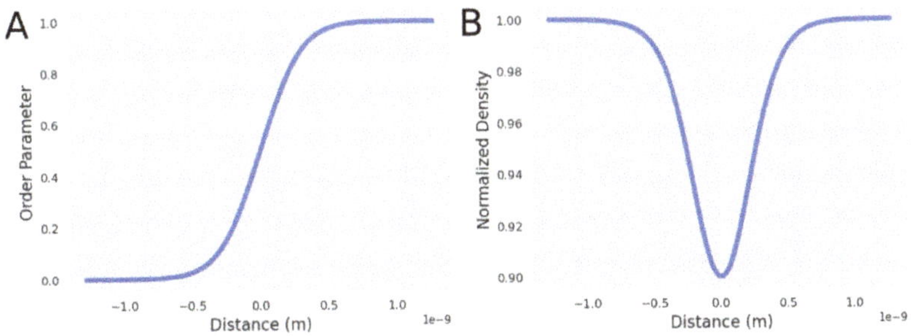


**Figure 3.** Plot (**A**): the equilibrium order parameter field $\phi$ achieved by minimizing a free energy functional with a volumetric free energy of the form given by Equation (22). Plot (**B**): the equilibrium density profile obtained by taking the curve of $\phi$ present in plot (**A**) and using Equation (16) to determine $\rho$.

## 3. Methods

We seek to demonstrate the validity of coupling $\rho$ and $\phi$ in addition to the enhanced functionality afforded by doing so. To this end both one and two dimensional DPF models were developed that use explicit finite differencing to solve the model A equation for the time evolution of the density and order parameter fields. This section outlines a series of tests that are used to validate the use of coupled DPF simulations as well as demonstrate their ability to accurately describe grain boundary physics.

### *3.1. Equilibrium Grain Boundary Properties*

The two main equilibrium quantities of interest in grain boundary simulations are the boundary width and excess energy. Real grain boundaries are approximately 1 nm wide with excess free energies on the order of 1 $\frac{J}{m^2}$ at room temperature. These are average values under normal conditions with variations possible because of material composition, grain boundary type, temperature, stress state, etc. We evaluate the fitness of the coupled DPF model by varying the values of $\rho_{min}$ and $\kappa_\phi$ in order to see their influence on both grain boundary width and excess energy. The grain boundary energy is calculated using Equation (23).

$$\gamma = \frac{1}{A} \int [F_v - \rho F_v(\rho = 1)] dV \tag{23}$$

The grain boundary width can be calculated from the profile of $\rho$ using a simple threshold criterion, e.g., the grain boundary region consists of the region of the density curve where $\rho < \rho_{cut}$). The expression used for $\rho_{cut}$ is provided below.

$$\rho_{cut} = 1 - 0.01 \left( \frac{1 - \rho^0_{min}}{1 - \rho_{min}} \right) \tag{24}$$

### 3.2. Dynamic Properties of Planar Grain Boundaries

For planar grain boundaries, the intrinsic driving force for grain growth is zero. A number of non-intrinsic driving forces for grain growth can be imposed on planar grain boundaries such that grain boundary motion occurs. A linear relationship exists between the boundary velocity and the driving force per unit area (pressure) as shown in Equation (25) where $v$ equals the grain boundary velocity, $m_{GB}$ equals the grain boundary mobility, and $P$ equals the driving pressure for motion.

$$V_{GB} = m_{GB}P \tag{25}$$

In order for a phase-field model to be considered valid, Equation (25) should hold. More explicitly, the relationship between grain boundary velocity and driving force should be linear. The means by which an applied driving force can be exerted on a planar grain boundary in a phase-field simulation is shown below.

The relation between grain boundary velocity and the evolution of the field variable $\phi$ is given by Equation (26).

$$v_{GB} = \dot{\phi}|\nabla\phi|^{-1} \tag{26}$$

Combining Equations (25) and (26), one can express the evolution of the order parameter field in terms of the applied driving force.

$$\dot{\phi} = m_{GB}P|\nabla\phi| \tag{27}$$

The grain boundary mobility multiplied by the applied driving pressure we refer to as the speed factor. Finally, Equation (27) can be added to the model A equation (Equation (17) to provide a motion equation for the order parameter field that includes the influence of an external driving force for grain growth.

$$\dot{\phi} = -m_{\phi}\mu_{\phi} + m_{GB}|\nabla\phi|P \tag{28}$$

### 3.3. The Shrinking Circular Grain Problem

The shrinking circular grain problem is a useful means of evaluating the accuracy of phase-field methods because it can be compared with an analytical solution. We describe the mathematics surrounding the problem here as well as the method by which the order parameter mobility can be determined using the shrinking circular grain problem. The velocity of a grain boundary can be expressed as a mobility times a pressure.

$$V_{GB} = m_{GB}P \tag{29}$$

If we consider the case of a 2D circular grain with radius $r$, the intrinsic driving pressure can be expressed in terms of the grain boundary energy and the grain radius.

$$P_{intrinsic} = \frac{2\gamma}{r} \tag{30}$$

For a grain boundary that is moving with constant velocity parallel to the grain boundary surface normal, the forward motion of the grain boundary can be related to the local rate of change in the order parameter $\phi$ using the equations below.

$$V_{GB} = \frac{\dot{\phi}}{|\nabla\phi|} \tag{31}$$

$$\frac{2m_{GB}\gamma}{r} = -\frac{1}{|\nabla\phi|}(m_{\phi}\Delta\mu_{\phi}) \tag{32}$$

The mobility of the $\phi$ field can be obtained by rearranging this equation into the following form.

$$m_\phi = -\frac{2m_{GB}\gamma|\nabla\phi|}{r\Delta\mu_\phi} \tag{33}$$

The grain boundary mobility and excess free energy can be determined using molecular dynamics [24–28]. At the grain boundary center ($\phi = 0.5$), the bulk and reference terms of the density potential difference are zero such that $\Delta\mu_\phi = -2\kappa_\phi\nabla^2\phi$. In order to solve for $\nabla\phi$ and $\nabla^2\phi$, we assume the order parameter field can be accurately approximated using a logistic function.

$$\phi = \frac{1}{1+e^{-k(x-x_0)}} \tag{34}$$

In polar coordinates, the gradient and Laplacian of $\phi$ are

$$\nabla\phi = \hat{e}_r k\phi(1-\phi) \tag{35}$$

$$\nabla^2\phi = \frac{\partial^2\phi}{\partial r^2} + \frac{1}{r}\frac{\partial\phi}{\partial r} \tag{36}$$

If we evaluate the Laplacian of $\phi$ at $\phi = 0.5$ we obtain the following expression.

$$\nabla^2\phi = \frac{|\nabla|}{r} \tag{37}$$

Substituting back into Equation (33) we obtain the following relationship for the mobility of $\phi$.

$$m_\phi = \frac{2m_{GB}\gamma}{\kappa_\phi} \tag{38}$$

### 3.4. Free Energy Functional Parameterization

A number of parameters must be determined in order to simulate grain boundary motion through the minimization of Equation (1) with Equation (22) used as the volumetric free energy. All relevant parameters are listed in Table 1.

**Table 1.** Density phase-field simulation parameters.

| Symbol | Value | Description | Units |
|---|---|---|---|
| $E_{co}$ | $4.32 \times 10^5$ | The cohesive energy | $\frac{\text{J}}{\text{mole}}$ |
| $\hat{V}_0$ | $6.6 \times 10^{-6}$ | The equilibrium molar volume | $\frac{\text{m}^3}{\text{mole}}$ |
| $n$ | 4 | Free energy parameter | none |
| $a$ | $-4$ | Free energy parameter | none |
| $\kappa_\phi$ | $1.5 \times 10^{-14}$ | Gradient Energy coefficient | $\frac{\text{J}}{\text{m}}$ |
| $m_\phi$ | $6.5 \times 10^6$ | order parameter mobility | $\frac{1}{\text{Pas}}$ |
| $\rho_{min}$ | 0.9 | minimum grain boundary density | none |

The cohesive energy energy and equilibrium molar volume correspond to nickel. Free energy parameters were chosen because they satisfy satisfy thermodynamic criteria 1–5 and give rise to ideally shaped free energy curves. A "typical" value of $\rho_{min}$ was selected for grain boundaries studied in the Olmsted database and does not correspond to a specific grain boundary [29,30]. The value of $\kappa_\phi$ was set so that in combination with the other thermodynamic criteria the grain boundary width would equal one. The order parameter mobility was calculated using Equation (38). We assume that the grain boundary mobility is $100\frac{\text{m}}{\text{sGPa}}$. In the same vein as for the choice of $\rho_{min}$, a mobility of $100\frac{\text{m}}{\text{sGPa}}$ is a typical values for nickel grain boundaries as calculated in the study by Olmsted [29,30]. The authors emphasize here that the good agreement shown between our work (see results section) and the work performed by Olmsted is indicative that our modifications to the DPF model are valid.

## 4. Results

### 4.1. Equilibrium Results

The equilibrium one dimensional order parameter and density profiles associated with the coupled DPF free energy functional are presented in Figure 3. As mentioned in the theory section, the density profiles associated with the coupled form of the volumetric free energy function results in more accurate density profiles in comparison with atomistics than does the non-coupled volumetric free energy functional. In particular, the discontinuity in the first derivative of the density curve is eliminated at the grain boundary center without the "shoulder regions" and higher computational overhead that results from the inclusion of higher-order gradient energy terms.

If the gradient energy series is truncated to a single term, the relationship between grain boundary width and the gradient energy coefficient should follow a square root relationship [13]. Figure 4 demonstrates that such a relationship is obeyed for physically relevant grain boundary widths.

### 4.2. Dynamic Results

Using the coupled free energy expression, the density description of grain boundaries can be made mobile. For the steady state case of a grain boundary moving at constant velocity, Equation (25) has been shown to hold both experimentally and using molecular dynamic simulations [29,30]. Figure 5 illustrates that coupled density phase-field simulations also obey this trend.

For transient problems, we use the classic 2D phase-field problem of a shrinking circular grain to study the motion of coupled DPF boundaries. The analytical relation between the grain radius and time is given by Equation (39) [5].

$$r = \sqrt{r_0^2 - 4m_\phi\kappa_\phi t} \tag{39}$$

In Figure 6 one can see the comparison of the analytical solution with the numerical solution of a shrinking 4 nm grain. The numerical solution is nearly identical to the analytical solution, expressing both the correct magnitude and trend of the radius vs. time curve. Assumed in Equation (39) is that the radius of the spherical grain is appreciably larger than the width of the interface, thus the slight deviation of the numerical solution from the analytical solution for small r values is not surprising.

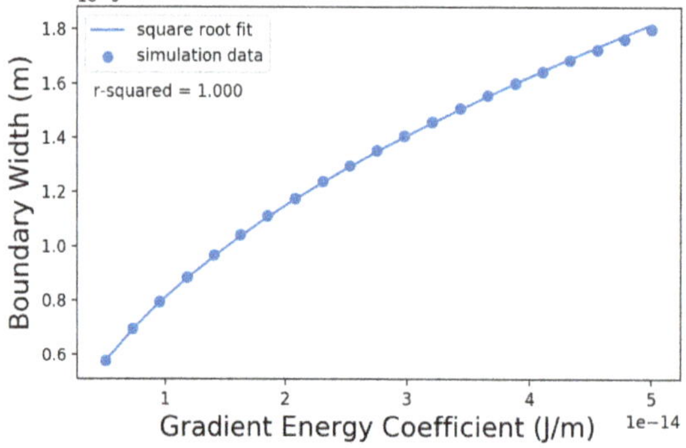


**Figure 4.** Plotted above is the equilibrium grain boundary width vs. the value of the gradient energy coefficient (data points) as well as a line of best fit corresponding to a square root function (the expected theoretical relationship). Visually the agreement between simulation and theory is exemplary and the $R^2$ value of the curve fit is 1 when calculated to three decimal places.

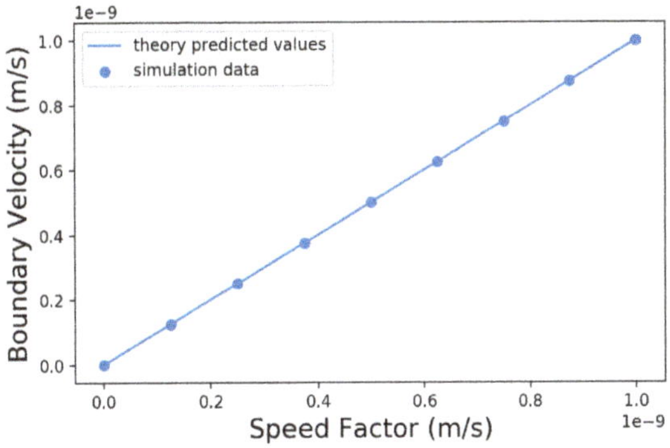


**Figure 5.** Plotted above are steady state grain boundary velocities achieved through the application of a synthetic driving force vs. speed factor (the theoretically predicted grain boundary velocity). It can be seen that there is good agreement between simulation and theory as indicated by the overlapping of the simulation data points and the solid line that represents the linear relationship between velocity and the product of grain boundary mobility and applied driving pressure.

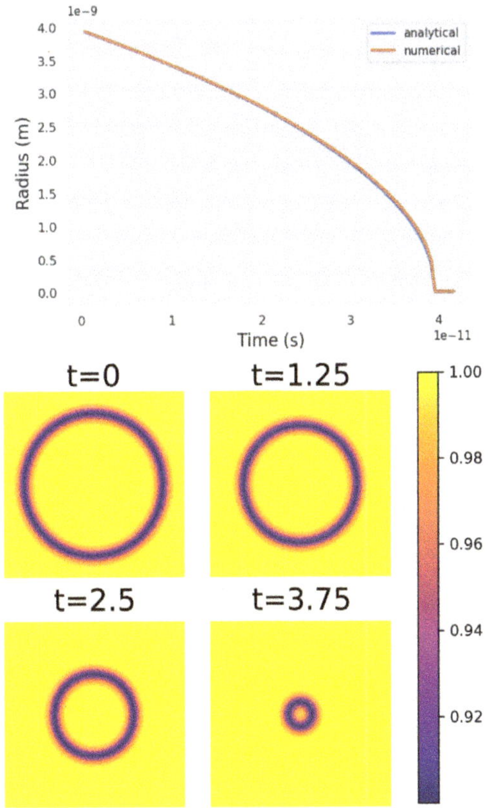


**Figure 6.** The shrinking circular grain problem modelled using the coupled DPF method. Note that the radius vs. time plot exhibits the characteristic parabolic shape. The color scale corresponds to the normalized density field.

## 5. Discussion

With respect to the derivation presented in Section 2.1, the resulting free energy functional has more utility than just meeting the thermodynamic consistency criteria. The well shape of the free energy curves presented in Figure 2 reflects the underlying atomic interactions that give rise to similarly shaped interatomic potential functions, where such a well shape motivated Jacobson to base a DPF free energy functional directly off of interatomic potentials, albeit being theoretically satisfying but computationally expensive. This computational expense is now avoided by using the traditional phase field order parameter constructed with the grain boundary center-line constraints. Equation (13) can easily be manipulated through the $a$ and $n$ parameters to achieve a well shape very nearly identical to those generated through interatomic potentials without the associated computational overhead. Furthermore, Equation (13) can be extended through the inclusion of more polynomial terms such that additional modification to the well shape can be achieved. As a result, the correct parameterization of the free energy functional gives rise to an accurate description of grain boundary physics in addition to some macroscale material properties such as bulk modulus.

The primary benefits of coupling the density field with an order parameter is that it provides a means by which to carry out dynamic field simulations in a more computationally efficient manner. An additional advantage that coupled DPF simulations have over "regular" DPF simulations is that expressing $\rho$ in terms of $\phi$ makes it much easier to incorporate DPF methods into traditional phase-field solvers. This back compatibility with existing software will not only make implementation easier, but will also allow for methods that have been used to accelerate classical phase-field simulations to be used to accelerate DPF simulations.

## 6. Conclusions

A set of thermodynamic criteria have been developed that can be used to ensure thermodynamic equilibrium and stability of the bulk state in density-based free energy functionals. The density field variable has been expressed in terms of traditional order parameter type variables in order to make DPF grain boundaries mobile. Approximating the density gradient energy sum in terms of $\phi$ results in a multiple order of magnitude increase in computational performance as well as a more accurate density profile. Finally, the dynamic DPF simulations are shown to be physical by simulating a circular grain whose change in radius with time matches the parabolic analytical solution. The combination of a more accurate density profile and the excellent agreement between the numerical and analytical solutions of the circular grain shrinkage problem indicate that coupling density fields with traditional order parameters in DPF simulations is a sound means by which to make DPF simulations dynamic while also improving simulation performance and accuracy.

**Author Contributions:** D.J. provided conceptualization, methodology, programing, validation, formal analysis, investigation, data curation, and writing the original draft. R.D.K. provided conceptualization, methodology, and review and editing. G.B.T. provided conceptualization, review and editing, supervision, project administration, and funding acquisition. All authors have read and agreed to the published version of the manuscript.

**Funding:** The authors thankfully acknowledge the National Science Foundation (DMR 1709803) for supporting this work. RDK acknowledges the financial support from DFG, project DA 1655/2-1 in the Heisenberg program.

**Data Availability Statement:** Data will be made available upon request to the authors.

**Conflicts of Interest:** The authors declare that they have no known competing financial interest or personal relationships that could have appeared to influence the work reported in this paper.

## Appendix A. Density Polynomial Derivation

### General

$$F_v = \frac{\rho}{\hat{V}}\hat{F} \quad \text{(A1)}$$

$$\hat{F}(\rho = 1) = \hat{E}_{co} \quad \text{(A2)}$$

### Criteria

1. $\hat{F}(\rho = 1) = E_{co}$
2. $\hat{F}(\rho = 0) = 0$
3. $\hat{F}(0 < \rho < 1) < 0$
4. $\frac{\partial \hat{F}}{\partial \rho}|_{\rho=1} = 0$
5. $\frac{\partial \hat{F}}{\partial \rho}|_{\rho=0} = 0$

Assume $E_{co} < 0$
Assume the following form of the molar free energy

$$\hat{F} = \rho^n f(\rho) E_{co} \quad \text{(A3)}$$

We can now determine a second order polynomial form of $f(\rho)$ that satisfies the criteria listed above assuming that $n >= 2$.

$$f(\rho) = a\rho^2 + b\rho + c \quad \text{(A4)}$$

$$f'(\rho) = 2a\rho + b \quad \text{(A5)}$$

$$f''(\rho) = 2a \quad \text{(A6)}$$

$$\begin{aligned}\frac{\partial \hat{F}}{\partial \rho} &= E_{co}(n\rho^{n-1} f(\rho) + \rho^n f'(\rho)) \\ &= E_{co}\rho^{n-1}(a\rho^2(n+2) + b\rho(n+1) + cn)\end{aligned} \quad \text{(A7)}$$

$$\frac{\partial^2 \hat{F}}{\partial \rho^2} = E_{co}\rho^{n-2}[(n+2)(n+1)a\rho^2 + (n+1)nb\rho + n(n-1)c] \quad \text{(A8)}$$

- Criteria 1: $1 = a + b + c$
  from criteria 4: $c = 1 + a + n$
- Criteria 2: guaranteed by $n >= 1$
- Criteria 3:
- Criteria 4: $n + 2a + b = 0$
  $b = -(n + 2a)$
- Criteria 5: guaranteed by $n >= 2$

## References


1. Gottstein, G.; Shvindlerman, L.S. *Grain Boundary Migration in Metals*, 2nd ed.; CRC Series in Materials Science and Technology; CRC Press: Boca Raton, FL, USA, 2009. [CrossRef]
2. Provatas, N.; Elder, K. *Phase-Field Methods in Materials Science and Engineering*; John Wiley & Sons: Hoboken, NJ, USA, 2011.
3. Steinbach, I. Phase-field models in materials science. *Model. Simul. Mater. Sci. Eng.* **2009**, 17, 073001. [CrossRef]
4. Chen, L.Q. Phase-Field Models for Microstructure Evolution. *Annu. Rev. Mater. Res.* **2002**, 32, 113–140. [CrossRef]
5. Darvishi Kamachali, R. Grain Boundary Motion in Polycrystalline Materials. Ph.D. Thesis, Ruhr-Universität Bochum, Bochum, Germany, 2012.
6. Kamachali, R.D.; Steinbach, I. 3-D phase-field simulation of grain growth: topological analysis versus mean-field approximations. *Acta Mater.* **2012**, 60, 2719–2728. [CrossRef]
7. Kamachali, R.D.; Abbondandolo, A.; Siburg, K.; Steinbach, I. Geometrical grounds of mean field solutions for normal grain growth. *Acta Mater.* **2015**, 90, 252–258. [CrossRef]

8. Kim, S.G.; Kim, D.I.; Kim, W.T.; Park, Y.B. Computer simulations of two-dimensional and three-dimensional ideal grain growth. *Phys. Rev. E* **2006**, *74*, 061605. [CrossRef] [PubMed]
9. Krill III, C.; Chen, L.Q. Computer simulation of 3-D grain growth using a phase-field model. *Acta Mater.* **2002**, *50*, 3059–3075. [CrossRef]
10. Miyoshi, E.; Takaki, T.; Ohno, M.; Shibuta, Y.; Sakane, S.; Shimokawabe, T.; Aoki, T. Ultra-large-scale phase-field simulation study of ideal grain growth. *npj Comput. Mater.* **2017**, *3*, 25. [CrossRef]
11. Steinbach, I.; Shchyglo, O. Phase-field modelling of microstructure evolution in solids: Perspectives and challenges. *Curr. Opin. Solid State Mater. Sci.* **2011**, *15*, 87–92. [CrossRef]
12. Chen, L.Q.; Wang, Y. The continuum field approach to modeling microstructural evolution. *JOM* **1996**, *48*, 13–18. [CrossRef]
13. Darvishi Kamachali, R. A model for grain boundary thermodynamics. *RSC Adv.* **2020**, *10*, 26728–26741.
14. Darvishi Kamachali, R.; Kwiatkowski da Silva, A.; McEniry, E.; Ponge, D.; Gault, B.; Neugebauer, J.; Raabe, D. Segregation-assisted spinodal and transient spinodal phase separation at grain boundaries. *npj Comput. Mater.* **2020**, *6*, 191.
15. Zhou, X.; Darvishi Kamachali, R.; Boyce, B.L.; Clark, B.G.; Raabe, D.; Thompson, G.B. Spinodal Decomposition in Nanocrystalline Alloys. *Acta Mater.* **2021**, *215*, 117054. [CrossRef]
16. Li, L.; Darvishi Kamachali, R.; Li, Z.; Zhang, Z. Grain boundary energy effect on grain boundary segregation in an equiatomic high-entropy alloy. *Phys. Rev. Mater.* **2020**, *4*, 053603. [CrossRef]
17. Wang, L.; Darvishi Kamachali, R. Incorporating elasticity into CALPHAD-informed density-based grain boundary phase diagrams reveals segregation transition in Al-Cu and Al-Cu-Mg alloys. *Comput. Mater. Sci.* **2021**, *199*, 110717. [CrossRef]
18. Wallis, T.; Kamachali, R.D. Grain boundary structural variations amplify segregation transition and stabilize co-existing spinodal interfacial phases. *Acta Mater.* **2023**, *242*, 118446. [CrossRef]
19. Wang, L.; Darvishi Kamachali, R. Density-based grain boundary phase diagrams: Application to Fe-Mn-Cr, Fe-Mn-Ni, Fe-Mn-Co, Fe-Cr-Ni and Fe-Cr-Co alloy systems. *Acta Mater.* **2021**, *207*, 116668. [CrossRef]
20. Wang, L.; Darvishi Kamachali, R. CALPHAD integrated grain boundary co-segregation design: Towards safe high-entropy alloys. *J. Alloys Compd.* **2023**, *933*, 167717. [CrossRef]
21. Jacobson, D.; Darvishi Kamachali, R.; Thompson, G.B. Molecular Phase Field: Coupling Density Phase Field Model with Atomistic Potentials. *Under Consideration for Publication in Computational Materials Science.*
22. Uberuaga, B.P.; Vernon, L.J.; Martinez, E.; Voter, A.F. The relationship between grain boundary structure, defect mobility, and grain boundary sink efficiency. *Sci. Rep.* **2015**, *5*, 9095. [CrossRef]
23. Philippe, T. Corners in phase-field theory. *Phys. Rev. E* **2021**, *103*, 032801. [CrossRef]
24. Tschopp, M.A.; McDowell, D.L. Structures and energies of Σ3 asymmetric tilt grain boundaries in copper and aluminium. *Philos. Mag.* **2007**, *87*, 3147–3173. [CrossRef]
25. Janssens, M.; Brett, J.M. Cultural intelligence in global teams: A fusion model of collaboration. *Group Organ. Manag.* **2006**, *31*, 124–153. [CrossRef]
26. Trautt, Z.T.; Upmanyu, M.; Karma, A. Interface mobility from interface random walk. *Science* **2006**, *314*, 632–635. [CrossRef]
27. Foiles, S.M.; Hoyt, J.J. Computation of grain boundary stiffness and mobility from boundary fluctuations. *Acta Mater.* **2006**, *54*, 3351–3357. [CrossRef]
28. Homer, E.R.; Holm, E.A.; Foiles, S.M.; Olmsted, D.L. Trends in grain boundary mobility: Survey of motion mechanisms. *JOM* **2014**, *66*, 114–120. [CrossRef]
29. Olmsted, D.L.; Foiles, S.M.; Holm, E.A. Survey of computed grain boundary properties in face-centered cubic metals: I. Grain boundary energy. *Acta Mater.* **2009**, *57*, 3694–3703. [CrossRef]
30. Olmsted, D.L.; Holm, E.A.; Foiles, S.M. Survey of computed grain boundary properties in face-centered cubic metals-II: Grain boundary mobility. *Acta Mater.* **2009**, *57*, 3704–3713. [CrossRef]

MDPI AG
Grosspeteranlage 5
4052 Basel
Switzerland
Tel.: +41 61 683 77 34

*Metals* Editorial Office
E-mail: metals@mdpi.com
www.mdpi.com/journal/metals

www.ingramcontent.com/pod-product-compliance
Lightning Source LLC
Chambersburg PA
CBHW041953260926
51850CB00019B/2697
* 9 7 8 3 7 2 5 8 7 3 6 8 5 *